Let's Keep in Touch

Follow Us

Visit US at

www.EffortlessMath.com

Online

f https://www.facebook.com/Effortlessmath

https://goo.gl/2B6qWW

Online Math Lessons

It's easy! Here's how it works.

1- Request a FREE introductory session.

2- Meet a Math tutor online.

3- Start Learning Math in Minutes.

Send Email to: info@EffortlessMath.com

www.EffortlessMath.com

... So Much More Online!

- FREE Math lessons

- More Math learning books!

- Online Math Tutors

Looking for an Online Math Tutor?

Need a PDF format of this book?

Send Email to: info@EffortlessMath.com

TSI Mathematics Workbook

2018 - 2019

The Most Comprehensive Review for the Math Section of the TSI TEST

By

Reza Nazari & Ava Ross

Copyright © 2018

Reza Nazari & Ava Ross

All rights reserved. No part of this publication may be reproduced, stored in a retrieval system, or transmitted in any form or by any means, electronic, mechanical, photocopying, recording, scanning, or otherwise, except as permitted under Section 107 or 108 of the 1976 United States Copyright Ac, without permission of the author.

All inquiries should be addressed to:

info@effortlessMath.com

www.EffortlessMath.com

ISBN-13: 978-1984924544

ISBN-10: 1984924540

Published by: Effortless Math Education

www.EffortlessMath.com

Description

Effortless Math TSI Workbook provides students with the confidence and math skills they need to succeed on the TSI Math, providing a solid foundation of basic Math topics with abundant exercises for each topic. It is designed to address the needs of TSI test takers who must have a working knowledge of basic Math.

This comprehensive workbook with over 2,500 sample questions and 2 complete TSI tests is all you need to fully prepare for the TSI Math. It will help you learn everything you need to ace the math section of the TSI.

There are more than 2,500 Math problems with answers in this book.

Effortless Math unique study program provides you with an in–depth focus on the math portion of the exam, helping you master the math skills that students find the most troublesome.

This workbook contains most common sample questions that are most likely to appear in the mathematics section of the TSI.

Inside the pages of this comprehensive Workbook, students can learn basic math operations in a structured manner with a complete study program to help them understand essential math skills. It also has many exciting features, including:

- Dynamic design and easy–to–follow activities
- A fun, interactive and concrete learning process
- Targeted, skill–building practices
- Fun exercises that build confidence
- Math topics are grouped by category, so you can focus on the topics you struggle on
- All solutions for the exercises are included, so you will always find the answers
- 2 Complete TSI Math Practice Tests that reflect the format and question types on TSI

Effortless Math TSI Workbook is an incredibly useful tool for those who want to review all topics being covered on the TSI test. It efficiently and effectively reinforces learning outcomes through engaging questions and repeated practice, helping you to quickly master basic Math skills.

About the Author

Reza Nazari is the author of more than 100 Math learning books including:
– **Math and Critical Thinking Challenges:** For the Middle and High School Student
– **ACT Math in 30 Days.**
– **ASVAB Math Workbook 2018 – 2019**
– **Effortless Math Education Workbooks**
– **and many more Mathematics books ...**

Reza is also an experienced Math instructor and a test–prep expert who has been tutoring students since 2008. Reza is the founder of Effortless Math Education, a tutoring company that has helped many students raise their standardized test scores—and attend the colleges of their dreams. Reza provides an individualized custom learning plan and the personalized attention that makes a difference in how students view math.

To ask questions about Math, you can contact Reza via email at:
reza@EffortlessMath.com

Find Reza's professional profile at:
goo.gl/zoC9rJ

Contents

Chapter 1: Fractions and Decimals

Math Topics that you'll learn today:

- ✓ Simplifying Fractions
- ✓ Adding and Subtracting Fractions
- ✓ Multiplying and Dividing Fractions
- ✓ Adding Mixed Numbers
- ✓ Subtract Mixed Numbers
- ✓ Multiplying Mixed Numbers
- ✓ Dividing Mixed Numbers
- ✓ Comparing Decimals
- ✓ Rounding Decimals

- ✓ Adding and Subtracting Decimals
- ✓ Multiplying and Dividing Decimals
- ✓ Converting Between Fractions, Decimals and Mixed Numbers
- ✓ Factoring Numbers
- ✓ Greatest Common Factor
- ✓ Least Common Multiple
- ✓ Divisibility Rules

"A Man is like a fraction whose numerator is what he is and whose denominator is what he thinks of himself. The larger the denominator, the smaller the fraction." –Tolstoy

Simplifying Fractions

✎ *Simplify the fractions.*

1) $\frac{22}{36}$ = $\frac{11}{18}$

2) $\frac{8}{10}$ ÷ $\frac{2}{2}$ = $\frac{4}{5}$

3) $\frac{12}{18}$ ÷ $\frac{2}{2}$ = $\frac{6}{9}$ ÷ $\frac{3}{3}$ = $\frac{2}{3}$

4) $\frac{6}{8}$ ÷ $\frac{2}{2}$ = $\frac{3}{4}$

5) $\frac{13}{39}$ ÷ $\frac{13}{13}$ = $\frac{1}{3}$

6) $\frac{5}{20}$ ÷ $\frac{5}{5}$ = $\frac{1}{4}$

7) $\frac{16}{36}$

8) $\frac{18}{36}$

9) $\frac{20}{50}$ ÷ $\frac{10}{10}$ = $\frac{2}{5}$

10) $\frac{6}{54}$ ÷ $\frac{6}{6}$ = $\frac{1}{9}$

11) $\frac{45}{81}$

12) $\frac{21}{28}$

13) $\frac{35}{56}$

14) $\frac{52}{64}$

15) $\frac{13}{65}$

16) $\frac{44}{77}$ ÷ $\frac{11}{11}$ = $\frac{4}{7}$

17) $\frac{21}{42}$

18) $\frac{15}{36}$

19) $\frac{9}{24}$ ÷ $\frac{3}{3}$ = $\frac{3}{8}$

20) $\frac{20}{80}$ ÷ $\frac{10}{10}$ = $\frac{2}{8}$ ÷ $\frac{2}{2}$ = $\frac{1}{4}$

21) $\frac{25}{45}$ ÷ $\frac{5}{5}$ = $\frac{5}{9}$

Adding and Subtracting Fractions

✍ **Add fractions.**

1) $\dfrac{2}{3} + \dfrac{1}{2}$

4) $\dfrac{7}{4} + \dfrac{5}{9}$

7) $\dfrac{3}{4} + \dfrac{2}{5}$

2) $\dfrac{3}{5} + \dfrac{1}{3}$

5) $\dfrac{2}{5} + \dfrac{1}{5}$

8) $\dfrac{2}{3} + \dfrac{1}{5}$

3) $\dfrac{5}{6} + \dfrac{1}{2}$

6) $\dfrac{3}{7} + \dfrac{1}{2}$

9) $\dfrac{16}{25} + \dfrac{3}{5}$

✍ **Subtract fractions.**

10) $\dfrac{4}{5} - \dfrac{2}{5}$

13) $\dfrac{8}{9} - \dfrac{3}{5}$

16) $\dfrac{3}{4} - \dfrac{13}{18}$

11) $\dfrac{3}{5} - \dfrac{2}{7}$

14) $\dfrac{3}{7} - \dfrac{3}{14}$

17) $\dfrac{5}{8} - \dfrac{2}{5}$

12) $\dfrac{1}{2} - \dfrac{1}{3}$

15) $\dfrac{4}{15} - \dfrac{1}{10}$

18) $\dfrac{1}{2} - \dfrac{1}{9}$

Multiplying and Dividing Fractions

✎ *Multiplying fractions. Then simplify.*

1) $\dfrac{1}{5} \times \dfrac{2}{3}$

2) $\dfrac{3}{4} \times \dfrac{2}{3}$

3) $\dfrac{2}{5} \times \dfrac{3}{7}$

4) $\dfrac{3}{8} \times \dfrac{1}{3}$

5) $\dfrac{3}{5} \times \dfrac{2}{5}$

6) $\dfrac{7}{9} \times \dfrac{1}{3}$

7) $\dfrac{2}{3} \times \dfrac{3}{8}$

8) $\dfrac{1}{4} \times \dfrac{1}{3}$

9) $\dfrac{5}{7} \times \dfrac{7}{12}$

✎ *Dividing fractions.*

10) $\dfrac{2}{9} \div \dfrac{1}{4}$

11) $\dfrac{1}{2} \div \dfrac{1}{3}$

12) $\dfrac{6}{11} \div \dfrac{3}{4}$

13) $\dfrac{11}{14} \div \dfrac{1}{10}$

14) $\dfrac{3}{5} \div \dfrac{5}{9}$

15) $\dfrac{1}{2} \div \dfrac{1}{2}$

16) $\dfrac{3}{5} \div \dfrac{1}{5}$

17) $\dfrac{12}{21} \div \dfrac{3}{7}$

18) $\dfrac{5}{14} \div \dfrac{9}{10}$

Adding Mixed Numbers

✎*Add.*

1) $4\frac{1}{2} + 5\frac{1}{2}$

2) $2\frac{3}{8} + 3\frac{1}{8}$

3) $6\frac{1}{5} + 3\frac{2}{5}$

4) $1\frac{1}{3} + 2\frac{2}{3}$

5) $5\frac{1}{6} + 5\frac{1}{2}$

6) $3\frac{1}{3} + 1\frac{1}{3}$

7) $1\frac{10}{11} + 1\frac{1}{3}$

8) $2\frac{3}{6} + 1\frac{1}{2}$

9) $5\frac{3}{5} + 5\frac{1}{5}$

10) $7 + \frac{1}{5}$

11) $1\frac{5}{7} + \frac{1}{3}$

12) $2\frac{1}{4} + 1\frac{1}{2}$

Subtract Mixed Numbers

✍ *Subtract.*

1) $4\frac{1}{2} - 3\frac{1}{2}$

2) $3\frac{3}{8} - 3\frac{1}{8}$

3) $6\frac{3}{5} - 5\frac{1}{5}$

4) $2\frac{1}{3} - 1\frac{2}{3}$

5) $6\frac{1}{6} - 5\frac{1}{2}$

6) $3\frac{1}{3} - 1\frac{1}{3}$

7) $2\frac{10}{11} - 1\frac{1}{3}$

8) $2\frac{1}{2} - 1\frac{1}{2}$

9) $6\frac{3}{5} - 2\frac{1}{5}$

10) $7\frac{2}{5} - 1\frac{1}{5}$

11) $2\frac{5}{7} - 1\frac{1}{3}$

12) $2\frac{1}{4} - 1\frac{1}{2}$

Multiplying Mixed Numbers

✎*Find each product.*

1) $1\frac{2}{3} \times 1\frac{1}{4}$

2) $1\frac{3}{5} \times 1\frac{2}{3}$

3) $1\frac{2}{3} \times 3\frac{2}{7}$

4) $4\frac{1}{8} \times 1\frac{2}{5}$

5) $2\frac{2}{5} \times 3\frac{1}{5}$

6) $1\frac{1}{3} \times 1\frac{2}{3}$

7) $1\frac{5}{8} \times 2\frac{1}{2}$

8) $3\frac{2}{5} \times 2\frac{1}{5}$

9) $2\frac{2}{3} \times 4\frac{1}{4}$

10) $2\frac{3}{5} \times 1\frac{2}{4}$

11) $1\frac{1}{3} \times 1\frac{1}{4}$

12) $3\frac{2}{5} \times 1\frac{1}{5}$

Dividing Mixed Numbers

✍ *Find each quotient.*

1) $2\frac{1}{5} \div 2\frac{1}{2}$

2) $2\frac{3}{5} \div 1\frac{1}{3}$

3) $3\frac{1}{6} \div 4\frac{2}{3}$

4) $1\frac{2}{3} \div 3\frac{1}{3}$

5) $4\frac{1}{8} \div 2\frac{2}{4}$

6) $3\frac{1}{2} \div 2\frac{3}{5}$

7) $3\frac{5}{9} \div 1\frac{2}{5}$

8) $2\frac{2}{7} \div 1\frac{1}{2}$

9) $3\frac{1}{5} \div 1\frac{1}{2}$

10) $4\frac{3}{5} \div 2\frac{1}{3}$

11) $6\frac{1}{6} \div 1\frac{2}{3}$

12) $2\frac{2}{3} \div 1\frac{1}{3}$

Comparing Decimals

✏️ *Write the correct comparison symbol (>, < or =).*

1) 1.25   2.3

2) 0.5   0.23

3) 3.2   3.2

4) 4.58   45.8

5) 2.75   0.275

6) 5.2   5

7) 3.1   0.31

8) 6.33   0.733

9) 8   0.8

10) 4.56 0.456

11) 1.12 1.14

12) 2.77 2.78

13) 6.08 6.11

14) 1.11 0.211

15) 2.6   2.55

16) 1.24 1.25

17) 5.52 0.552

18) 0.33 0.033

19) 14.4 14.4

20) 0.05 0.50

21) 0.59 0.7

22) 0.5   0.05

23) 0.90 0.9

24) 0.27 0.4

Rounding Decimals

✍️ *Round each decimal number to the nearest place indicated.*

1) 0.2̲3

2) 4.0̲4

3) 5.6̲23

4) 0.2̲66

5) 6̲.37

6) 0.8̲8

7) 8.2̲4

8) 7̲.0760

9) 1.6̲29

10) 6.3̲959

11) 1̲.9

12) 5̲.2167

13) 5.8̲63

14) 8.5̲4

15) 80̲.69

16) 65̲.85

17) 70.7̲8

18) 615̲.755

19) 16̲.4

20) 95̲.81

21) 2̲.408

22) 76̲.3

23) 116.5̲14

24) 8.0̲6

Adding and Subtracting Decimals

✍ *Add and subtract decimals.*

1) $\begin{array}{r} 15.14 \\ -\ 12.18 \\ \hline \end{array}$

3) $\begin{array}{r} 82.56 \\ +\ 12.28 \\ \hline \end{array}$

5) $\begin{array}{r} 90.37 \\ +\ 56.97 \\ \hline \end{array}$

2) $\begin{array}{r} 65.72 \\ +\ 43.67 \\ \hline \end{array}$

4) $\begin{array}{r} 34.18 \\ -\ 23.45 \\ \hline \end{array}$

6) $\begin{array}{r} 45.78 \\ -\ 23.39 \\ \hline \end{array}$

✍ *Solve.*

7) ____ + 1.3 = 4.8

10) 6.9 + ____ = 16.4

8) 4.2 + ____ = 11.6

11) ____ + 5.1 = 8.6

9) 9.9 + ____ = 16

12) ____ + 7.9 = 15.2

Multiplying and Dividing Decimals

🖎*Find each product.*

1)
$$
\begin{array}{r}
4.5 \\
\times\ 1.6 \\
\hline
\end{array}
$$

4)
$$
\begin{array}{r}
8.9 \\
\times\ 9.7 \\
\hline
\end{array}
$$

7)
$$
\begin{array}{r}
5.7 \\
\times\ 7.8 \\
\hline
\end{array}
$$

2)
$$
\begin{array}{r}
7.7 \\
\times\ 9.9 \\
\hline
\end{array}
$$

5)
$$
\begin{array}{r}
15.1 \\
\times\ 12.6 \\
\hline
\end{array}
$$

8)
$$
\begin{array}{r}
98.20 \\
\times\ 100 \\
\hline
\end{array}
$$

3)
$$
\begin{array}{r}
2.6 \\
\times\ 1.5 \\
\hline
\end{array}
$$

6)
$$
\begin{array}{r}
6.9 \\
\times\ 3.3 \\
\hline
\end{array}
$$

9)
$$
\begin{array}{r}
23.99 \\
\times\ 1000 \\
\hline
\end{array}
$$

🖎*Find each quotient.*

10) $9.2 \div 3.6$

13) $6.5 \div 8.1$

16) $4.24 \div 10$

11) $27.6 \div 3.8$

14) $1.4 \div 10$

17) $14.6 \div 100$

12) $12.6 \div 4.7$

15) $3.6 \div 100$

18) $1.8 \div 1000$

Converting Between Fractions, Decimals and Mixed Numbers

✍ *Convert fractions to decimals.*

1) $\dfrac{9}{10}$

4) $\dfrac{2}{5}$

7) $\dfrac{12}{10}$

2) $\dfrac{56}{100}$

5) $\dfrac{3}{9}$

8) $\dfrac{8}{5}$

3) $\dfrac{3}{4}$

6) $\dfrac{40}{50}$

9) $\dfrac{69}{10}$

✍ *Convert decimal into fraction or mixed numbers.*

10)   0.3

14)   0.8

18)   0.08

11)   4.5

15)   0.25

19)   0.45

12)   2.5

16)   0.14

20)   2.6

13)   2.3

17)   0.2

21)   5.2

Factoring Numbers

✐ *List all positive factors of each number.*

1) 68

2) 56

3) 24

4) 40

5) 86

6) 78

7) 50

8) 98

9) 45

10) 26

11) 54

12) 28

13) 55

14) 85

15) 48

✐ *List the prime factorization for each number.*

16) 50

17) 25

18) 69

19) 21

20) 45

21) 68

22) 26

23) 86

24) 93

Greatest Common Factor

✏️*Find the GCF for each number pair.*

1) 20, 30	8) 34, 6	15) 45, 8
2) 4, 14	9) 4, 10	16) 90, 35
3) 5, 45	10) 5, 3	17) 78, 34
4) 68, 12	11) 6, 16	18) 55, 75
5) 5, 12	12) 30, 3	19) 60, 72
6) 15, 27	13) 24, 28	20) 100, 78
7) 3, 24	14) 70, 10	21) 30, 40

Least Common Multiple

✍ *Find the LCM for each number pair.*

1) 4, 14

2) 5, 15

3) 16, 10

4) 4, 34

5) 8, 3

6) 12, 24

7) 9, 18

8) 5, 6

9) 8, 19

10) 9, 21

11) 19, 29

12) 7, 6

13) 25, 6

14) 4, 8

15) 30, 10, 50

16) 18, 36, 27

17) 12, 8, 18

18) 8, 18, 4

19) 26, 20, 30

20) 10, 4, 24

21) 15, 30, 45

Answers of Worksheets – Chapter 1

Simplifying Fractions

1) $\frac{11}{18}$

2) $\frac{4}{5}$

3) $\frac{2}{3}$

4) $\frac{3}{4}$

5) $\frac{1}{3}$

6) $\frac{1}{4}$

7) $\frac{4}{9}$

8) $\frac{1}{2}$

9) $\frac{2}{5}$

10) $\frac{1}{9}$

11) $\frac{5}{9}$

12) $\frac{3}{4}$

13) $\frac{5}{8}$

14) $\frac{13}{16}$

15) $\frac{1}{5}$

16) $\frac{4}{7}$

17) $\frac{1}{2}$

18) $\frac{5}{12}$

19) $\frac{3}{8}$

20) $\frac{1}{4}$

21) $\frac{5}{9}$

Adding and Subtracting Fractions

1) $\frac{7}{6}$

2) $\frac{14}{15}$

3) $\frac{4}{3}$

4) $\frac{83}{36}$

5) $\frac{3}{5}$

6) $\frac{13}{14}$

7) $\frac{23}{20}$

8) $\frac{13}{15}$

9) $\frac{31}{25}$

10) $\frac{2}{5}$

11) $\frac{11}{35}$

12) $\frac{1}{6}$

13) $\frac{13}{45}$

14) $\frac{3}{14}$

15) $\frac{1}{6}$

16) $\frac{1}{36}$

17) $\frac{9}{40}$

18) $\frac{7}{18}$

Multiplying and Dividing Fractions

1) $\frac{2}{15}$

2) $\frac{1}{2}$

3) $\frac{6}{35}$

4) $\frac{1}{8}$

5) $\frac{6}{25}$

6) $\frac{7}{27}$

7) $\frac{1}{4}$

8) $\frac{1}{12}$

9) $\frac{5}{12}$

10) $\frac{8}{9}$

11) $\frac{3}{2}$

12) $\frac{8}{11}$

13) $\frac{55}{7}$

14) $\frac{27}{25}$

15) 1

16) 3

17) $\frac{4}{3}$

18) $\frac{25}{63}$

Adding Mixed Numbers

1) 10

2) $5\frac{1}{2}$

3) $9\frac{3}{5}$

4) 4

5) $10\frac{2}{3}$

6) $4\frac{2}{3}$

7) $3\frac{8}{33}$

8) 4

9) $10\frac{4}{5}$

10) $7\frac{1}{5}$

11) $2\frac{1}{21}$

12) $3\frac{3}{4}$

Subtract Mixed Numbers

1) 1

2) $\frac{1}{4}$

3) $1\frac{2}{5}$

4) $\frac{2}{3}$

5) $\frac{2}{3}$

6) 2

7) $1\frac{19}{33}$

8) 1

9) $4\frac{2}{5}$

10) $6\frac{1}{5}$

11) $1\frac{8}{21}$

12) $\frac{3}{4}$

Multiplying Mixed Numbers

1) $2\frac{1}{12}$

2) $2\frac{2}{3}$

3) $5\frac{10}{21}$

4) $5\frac{31}{40}$

5) $7\frac{17}{25}$

6) $2\frac{2}{9}$

7) $4\frac{1}{16}$

8) $7\frac{12}{25}$

9) $11\frac{1}{3}$

10) $3\frac{9}{10}$

11) $1\frac{2}{3}$

12) $4\frac{2}{25}$

Dividing Mixed Numbers

1) $\frac{22}{25}$

2) $1\frac{19}{20}$

3) $\frac{19}{28}$

4) $\frac{1}{2}$

5) $1\frac{13}{20}$

6) $1\frac{9}{26}$

7) $2\frac{34}{63}$

8) $1\frac{11}{21}$

9) $2\frac{2}{15}$

10) $1\frac{34}{35}$

11) $3\frac{7}{10}$

12) 2

Comparing Decimals

1) $1.25 < 2.3$

2) $0.5 > 0.23$

3) $3.2 = 3.2$

4) $4.58 < 45.8$

5) $2.75 > 0.275$

6) $5.2 > 5$

7) $3.1 > 0.31$

8) $6.33 > 0.733$

9) $8 > 0.8$

10) $4.56 > 0.456$

11) $1.12 < 1.14$

12) $2.77 < 2.78$

13) $6.08 < 6.11$

14) $1.11 > 0.211$

15) $2.6 > 2.55$

16) $1.24 < 1.25$

17) $5.52 > 0.552$

18) $0.33 > 0.033$

19) $14.4 = 14.4$

20) $0.05 < 0.50$

21) $0.59 < 0.7$

22) $0.5 > 0.05$

23) $0.90 = 0.9$

24) $0.27 < 0.4$

Rounding Decimals

1) 0.2
2) 4.0
3) 5.6
4) 0.3
5) 6
6) 0.9
7) 8.2
8) 7

9) 1.63
10) 6.4
11) 2
12) 5
13) 5.9
14) 8.5
15) 81
16) 66

17) 70.8
18) 616
19) 16
20) 96
21) 2
22) 76
23) 116.5
24) 8.1

Adding and Subtracting Decimals

1) 2.96
2) 109.39
3) 94.84
4) 10.73

5) 147.34
6) 22.39
7) 3.5
8) 7.4

9) 6.1
10) 9.5
11) 3.5
12) 7.3

Multiplying and Dividing Decimals

1) 7.2
2) 76.23
3) 3.9
4) 86.33
5) 190.26
6) 22.77

7) 44.46
8) 9820
9) 23990
10) 2.5555…
11) 7.2631…
12) 2.6808…

13) 0.8024…
14) 0.14
15) 0.036
16) 0.424
17) 0.146
18) 0.0018

Converting Between Fractions, Decimals and Mixed Numbers

1) 0.9
2) 0.56
3) 0.75
4) 0.4
5) 0.333…
6) 0.8

7) 1.2
8) 1.6
9) 6.9
10) $\frac{3}{10}$
11) $4\frac{1}{2}$

12) $2\frac{1}{2}$
13) $2\frac{3}{10}$
14) $\frac{4}{5}$
15) $\frac{1}{4}$

16) $\frac{7}{50}$ 18) $\frac{2}{25}$ 20) $2\frac{3}{5}$

17) $\frac{1}{5}$ 19) $\frac{9}{20}$ 21) $5\frac{1}{5}$

Factoring Numbers

1) 1, 2, 4, 17, 34, 68
2) 1, 2, 4, 7, 8, 14, 28, 56
3) 1, 2, 3, 4, 6, 8, 12, 24
4) 1, 2, 4, 5, 8, 10, 20, 40
5) 1, 2, 43, 86
6) 1, 2, 3, 6, 13, 26, 39, 78
7) 1, 2, 5, 10, 25, 50
8) 1, 2, 7, 14, 49, 98
9) 1, 3, 5, 9, 15, 45
10) 1, 2, 13, 26
11) 1, 2, 3, 6, 9, 18, 27, 54
12) 1, 2, 4, 7, 14, 28

13) 1, 5, 11, 55
14) 1, 5, 17, 85
15) 1, 2, 3, 4, 6, 8, 12, 16, 24, 48
16) $2 \times 5 \times 5$
17) 5×5
18) 3×23
19) 3×7
20) $3 \times 3 \times 5$
21) $2 \times 2 \times 17$
22) 2×13
23) 2×43
24) 3×31

Greatest Common Factor

1) 10
2) 2
3) 5
4) 4
5) 1
6) 3
7) 3

8) 2
9) 2
10) 1
11) 2
12) 3
13) 4
14) 10

15) 1
16) 5
17) 2
18) 5
19) 12
20) 2
21) 10

Least Common Multiple

1) 28
2) 15
3) 80
4) 68
5) 24
6) 24
7) 18

8) 30
9) 152
10) 63
11) 551
12) 42
13) 150
14) 8

15) 150
16) 108
17) 72
18) 72
19) 780
20) 120
21) 90

Chapter 2: Real Numbers and Integers

Math Topics that you'll learn today:

- ✓ Adding and Subtracting Integers
- ✓ Multiplying and Dividing Integers
- ✓ Ordering Integers and Numbers
- ✓ Arrange and Order, Comparing Integers
- ✓ Order of Operations
- ✓ Mixed Integer Computations
- ✓ Integers and Absolute Value

"Wherever there is number, there is beauty." –Proclus

Adding and Subtracting Integers

✍ *Find the sum.*

1) $(-12) + (-4)$

2) $5 + (-24)$

3) $(-14) + 23$

4) $(-8) + (39)$

5) $43 + (-12)$

6) $(-23) + (-4) + 3$

7) $4 + (-12) + (-10) + (-25)$

8) $19 + (-15) + 25 + 11$

9) $(-9) + (-12) + (32 - 14)$

10) $4 + (-30) + (45 - 34)$

✍ *Find the difference.*

11) $(-14) - (-9) - (18)$

12) $(-9) - (-25)$

13) $(-12) - (8)$

14) $(28) - (-4)$

15) $(34) - (2)$

16) $(55) - (-5) + (-4)$

17) $(9) - (2) - (-5)$

18) $(2) - (4) - (-15)$

19) $(23) - (4) - (-34)$

20) $(-45) - (-87)$

Multiplying and Dividing Integers

✎ *Find each product.*

1) $(-8) \times (-2)$

2) 3×6

3) $(-4) \times 5 \times (-6)$

4) $2 \times (-6) \times (-6)$

5) $11 \times (-12)$

6) $10 \times (-5)$

7) 8×8

8) $(-8) \times (-9)$

9) $6 \times (-5) \times 3$

10) $6 \times (-1) \times 2$

✎ *Find each quotient.*

11) $18 \div 3$

12) $(-24) \div 4$

13) $(-63) \div (-9)$

14) $54 \div 9$

15) $20 \div (-2)$

16) $(-66) \div (-11)$

17) $64 \div 8$

18) $(-121) \div 11$

19) $72 \div 9$

20) $16 \div 4$

Ordering Integers and Numbers

✎ *Order each set of integers from least to greatest.*

1) − 15, − 19, 20, − 4, 1   ___, ___, ___, ___, ___, ___

2) 6, − 5, 4, − 3, 2   ___, ___, ___, ___, ___, ___

3) 15, − 42, 19, 0, − 22   ___, ___, ___, ___, ___, ___

4) 26, − 91, 0, − 13, 67, − 55   ___, ___, ___, ___, ___, ___

5) − 17, − 71, 90, − 25, − 54, − 39   ___, ___, ___, ___, ___, ___

6) 98, 5, 46, 19, 77, 24   ___, ___, ___, ___, ___, ___

✎ *Order each set of integers from greatest to least.*

7) − 2, 5, − 3, 6, − 4   ___, ___, ___, ___, ___, ___

8) − 37, 7, − 17, 27, 47   ___, ___, ___, ___, ___, ___

9) 32, − 27, 19, − 17, 15   ___, ___, ___, ___, ___, ___

10) 68, 81, 21, − 18, 94, 72   ___, ___, ___, ___, ___, ___

Arrange, Order, and Comparing Integers

✍ *Arrange these integers in descending order.*

1) $21, 71, -18, -10, 82$ ___, ___, ___, ___, ___, ___

2) $15, 11, 20, 12, -9, -5$ ___, ___, ___, ___, ___, ___

3) $-5, 20, 15, 9, -11$ ___, ___, ___, ___, ___, ___

4) $19, 18, -9, -6, -11$ ___, ___, ___, ___, ___, ___

5) $56, -34, -12, -5, 32$ ___, ___, ___, ___, ___, ___

✍ *Compare. Use >, =, <*

6) -8 ___ 12 11) -56 ___ -58

7) -10 ___ -16 12) 78 ___ 87

8) 43 ___ 34 13) -92 ___ -102

9) 15 ___ -16 14) -12 ___ -12

10) -354 ___ -345 15) -721 ___ -821

Order of Operations

✍️ *Evaluate each expression.*

1) $(2 \times 2) + 5$

2) $24 - (3 \times 3)$

3) $(6 \times 4) + 8$

4) $25 - (4 \times 2)$

5) $(6 \times 5) + 3$

6) $64 - (2 \times 4)$

7) $25 + (1 \times 8)$

8) $(6 \times 7) + 7$

9) $48 \div (4 + 4)$

10) $(7 + 11) \div (-2)$

11) $9 + (2 \times 5) + 10$

12) $(5 + 8) \times \dfrac{3}{5} + 2$

13) $2 \times 7 - \left(\dfrac{10}{9 - 4}\right)$

14) $(12 + 2 - 5) \times 7 - 1$

15) $\left(\dfrac{7}{5 - 1}\right) \times (2 + 6) \times 2$

16) $20 \div (4 - (10 - 8))$

17) $\dfrac{50}{4(5 - 4) - 3}$

18) $2 + (8 \times 2)$

Mixed Integer Computations

✎*Compute.*

1) $(-70) \div (-5)$

2) $(-14) \times 3$

3) $(-4) \times (-15)$

4) $(-65) \div 5$

5) $18 \times (-7)$

6) $(-12) \times (-2)$

7) $\dfrac{(-60)}{(-20)}$

8) $24 \div (-8)$

9) $22 \div (-11)$

10) $\dfrac{(-27)}{3}$

11) $4 \times (-4)$

12) $\dfrac{(-48)}{12}$

13) $(-14) \times (-2)$

14) $(-7) \times (7)$

15) $\dfrac{-30}{-6}$

16) $(-54) \div 6$

17) $(-60) \div (-5)$

18) $(-7) \times (-12)$

19) $(-14) \times 5$

20) $88 \div (-8)$

Integers and Absolute Value

✍ **Write absolute value of each number.**

1) -4

2) -7

3) -8

4) 4

5) 5

6) -10

7) 1

8) 6

9) 8

10) -2

11) -1

12) 10

13) 3

14) 7

15) -5

16) -3

17) -9

18) 2

19) 4

20) -6

21) 9

✍ **Evaluate.**

22) $|-43| - |12| + 10$

23) $76 + |-15 - 45| - |3|$

24) $30 + |-62| - 46$

25) $|32| - |-78| + 90$

26) $|-35 + 4| + 6 - 4$

27) $|-4| + |-11|$

28) $|-6 + 3 - 4| + |7 + 7|$

29) $|-9| + |-19| - 5$

Answers of Worksheets – Chapter 2

Adding and Subtracting Integers

1) − 16
2) − 19
3) 9
4) 31
5) 31
6) − 24
7) − 43

8) 40
9) − 3
10) − 15
11) − 23
12) 16
13) − 20
14) 32

15) 32
16) 56
17) 12
18) 13
19) 53
20) 42

Multiplying and Dividing Integers

1) 16
2) 18
3) 120
4) 72
5) − 132
6) − 50
7) 64

8) 72
9) − 90
10) − 12
11) 6
12) − 6
13) 7
14) 6

15) − 10
16) 6
17) 8
18) − 11
19) 8
20) 4

Ordering Integers and Numbers

1) − 19, − 15, − 4, 1, 20
2) − 5, − 3, 2, 4, 6
3) − 42, − 22, 0, 15, 19
4) − 91, − 55, − 13, 0, 26, 67
5) − 71, − 54, − 39, − 25, − 17, 90

6) 5, 19, 24, 46, 77, 98
7) 6, 5, − 2, − 3, − 4
8) 47, 27, 7, − 17, − 37
9) 32, 19, 15, − 17, − 27
10) 94, 81, 72, 68, 21, − 18

Arrange and Order, Comparing Integers

1) 82, 71, 21, − 10, − 18

2) 20, 15, 12, 11, − 5, − 9

3) 20, 15, 9, − 5, −11

4) 19, 18, − 6, − 9, − 11

5) 56, 32, − 5, − 12, − 34

6) <	10) <	14) =
7) >	11) >	15) >
8) >	12) <	
9) >	13) >	

Order of Operations

1) 9	7) 33	13) 12
2) 15	8) 49	14) 62
3) 32	9) 6	15) 28
4) 17	10) − 9	16) 10
5) 33	11) 29	17) 50
6) 56	12) 9.8	18) 18

Mixed Integer Computations

1) 14	8) − 3	15) 5
2) − 42	9) − 2	16) − 9
3) 60	10) − 9	17) 12
4) − 13	11) − 16	18) 84
5) − 126	12) − 4	19) − 70
6) 24	13) 28	20) − 11
7) 3	14) − 49	

Integers and Absolute Value

1) 4		11) 1		21) 9	
2) 7		12) 10		22) 41	
3) 8		13) 3		23) 133	
4) 4		14) 7		24) 46	
5) 5		15) 5		25) 44	
6) 10		16) 3		26) 33	
7) 1		17) 9		27) 15	
8) 6		18) 2		28) 21	
9) 8		19) 4		29) 23	
10) 2		20) 6			

Chapter 3: Proportions and Ratios

Math Topics that you'll learn today:

- ✓ Writing Ratios
- ✓ Simplifying Ratios
- ✓ Create a Proportion
- ✓ Similar Figures
- ✓ Simple Interest
- ✓ Ratio and Rates Word Problems

"Do not worry about your difficulties in mathematics. I can assure you mine are still ACTater." – Albert Einstein

Writing Ratios

✏️ *Express each ratio as a rate and unite rate.*

1) 120 miles on 4 gallons of gas.

2) 24 dollars for 6 books.

3) 200 miles on 14 gallons of gas

4) 24 inches of snow in 8 hours

✏️ *Express each ratio as a fraction in the simplest form.*

5) 3 feet out of 30 feet

6) 18 cakes out of 42 cakes

7) 16 dimes t0 24 dimes

8) 12 dimes out of 48 coins

9) 14 cups to 84 cups

10) 45 gallons to 65 gallons

11) 10 miles out of 40 miles

12) 22 blue cars out of 55 cars

13) 32 pennies to 300 pennies

14) 24 beetles out of 86 insects

Simplifying Ratios

✍ *Reduce each ratio.*

1) 21 : 49

2) 20 : 40

3) 10: 50

4) 14: 18

5) 45: 27

6) 49: 21

7) 100: 10

8) 12 : 8

9) 35 : 45

10) 8: 20

11) 25: 35

12) 21 : 27

13) 52 : 82

14) 12: 36

15) 24 : 3

16) 15: 30

17) 3 : 36

18) 8 : 16

19) 6 : 100

20) 2 : 20

21) 10: 60

22) 14: 63

23) 68: 80

24) 8: 80

Create a Proportion

✍️ *Create proportion from the given set of numbers.*

1) 1, 6, 2, 3

2) 12, 144, 1, 12

3) 16, 4, 8, 2

4) 9, 5, 27, 15

5) 7, 10, 60, 42

6) 8, 7, 24, 21

7) 10, 5, 8, 4

8) 3, 12, 8, 2

9) 2, 2, 1, 4

10) 3, 6, 7, 14

11) 2, 6, 5, 15

12) 7, 2, 14, 4

Similar Figures

✍ *Each pair of figures is similar. Find the missing side.*

1)

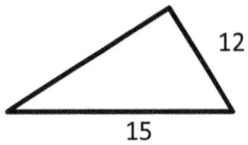 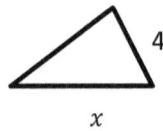

2)

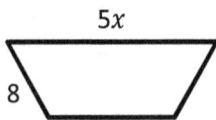 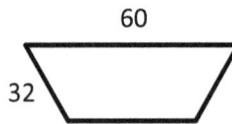

3)

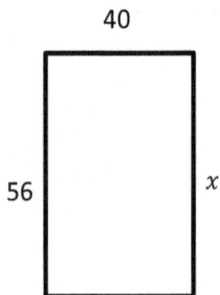 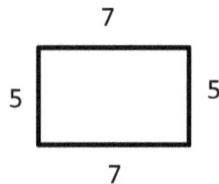

Simple Interest

✍ *Use simple interest to find the ending balance.*

1) $1,300 at 5% for 6 years.

2) $5,400 at 7.5% for 6 months.

3) $25,600 at 9.2% for 5 years

4) $24,000 at 8.5% for 9 years.

5) $450 at 7% for 8 years.

6) $54,200 at 8% for 5 years.

7) $240 interest is earned on a principal of $1500 at a simple interest rate of 4% interest per year. For how many years was the principal invested?

8) A new car, valued at $28,000, depreciates at 9% per year from original price. Find the value of the car 3 years after purchase.

9) Sara puts $2,000 into an investment yielding 5% annual simple interest; she left the money in for five years. How much interest does Sara get at the end of those five years?

Ratio and Rates Word Problems

✎*Solve.*

1) In a party, 10 soft drinks are required for every 12 guests. If there are 252 guests, how many soft drink is required?

2) In Jack's class, 18 of the students are tall and 10 are short. In Michael's class 54 students are tall and 30 students are short. Which class has a higher ratio of tall to short students?

3) Are these ratios equivalent?

   12 cards to 72 animals 11 marbles to 66 marbles

4) The price of 3 apples at the Quick Market is $1.44. The price of 5 of the same apples at Walmart is $2.50. Which place is the better buy?

5) The bakers at a Bakery can make 160 bagels in 4 hours. How many bagels can they bake in 16 hours? What is that rate per hour?

6) You can buy 5 cans of green beans at a supermarket for $3.40. How much does it cost to buy 35 cans of green beans?

Answers of Worksheets – Chapter 3

Writing Ratios

1) $\frac{120\ miles}{4\ gallons}$, 30 miles per gallon

2) $\frac{24\ dollars}{6\ books}$, 4.00 dollars per book

3) $\frac{200\ miles}{14\ gallons}$, 14.29 miles per gallon

4) $\frac{24"\ of\ snow}{8\ hours}$, 3 inches of snow per hour

5) $\frac{1}{10}$

6) $\frac{3}{7}$

7) $\frac{2}{3}$

8) $\frac{1}{4}$

9) $\frac{1}{6}$

10) $\frac{9}{13}$

11) $\frac{1}{4}$

12) $\frac{2}{5}$

13) $\frac{8}{75}$

14) $\frac{12}{43}$

Simplifying Ratios

1) 3 : 7
2) 1 : 2
3) 1 : 5
4) 7 : 9
5) 5 : 3
6) 7 : 3
7) 10 : 1
8) 3 : 2

9) 7 : 9
10) 2 : 5
11) 5 : 7
12) 7 : 9
13) 26 : 41
14) 1 : 3
15) 8 : 1
16) 1 : 2

17) 1 : 12
18) 1 : 2
19) 3 : 50
20) 1 : 10
21) 1 : 6
22) 2 : 9
23) 17 : 20
24) 1 : 10

Create a Proportion

1) 1 : 3 = 2 : 6

2) 12 : 144 = 1 : 12

3) 2 : 4 = 8 : 16

4) 5 : 15 = 9 : 27

5) 7 : 42, 10 : 60

6) 7 : 21 = 8 : 24

7) 8 : 10 = 4 : 5

8) 2 : 3 = 8 : 12

9) 4 : 2 = 2 : 1

10) 7 : 3 = 14 : 6   11) 5 : 2 = 15 : 6   12) 7 : 2 = 14 : 4

Similar Figures

1) 5   2) 3   3) 56

Simple Interest

1) $1,690.00   4) $42,360.00   7) 4 years

2) $5,602.50   5) $702.00   8) $20,440

3) $37,376.00   6) $75,880.00   9) $500

Ratio and Rates Word Problems

1) 210

2) The ratio for both class is equal to 9 to 5.

3) Yes! Both ratios are 1 to 6

4) The price at the Quick Market is a better buy.

5) 640, the rate is 40 per hour.

6) $23.80

Chapter 4: Percent

Math Topics that you'll learn today:

- ✓ Percentage Calculations
- ✓ Converting Between Percent, Fractions, and Decimals
- ✓ Percent Problems
- ✓ Markup, Discount, and Tax

"Do not worry about your difficulties in mathematics. I can assure you mine are still ACTater." – Albert Einstein

Percentage Calculations

✎ *Calculate the percentages.*

1) 50% of 25

2) 80% of 15

3) 30% of 34

4) 70% of 45

5) 10% of 0

6) 80% of 22

7) 65% of 8

8) 78% of 54

9) 50% of 80

10) 20% of 10

11) 40% of 40

12) 90% of 0

13) 20% of 70

14) 55% of 60

15) 80% of 10

16) 20% of 880

17) 70% of 100

18) 80% of 90

✎ *Solve.*

19) 50 is what percentage of 75?

20) What percentage of 100 is 70

21) Find what percentage of 60 is 35.

22) 40 is what percentage of 80?

Converting Between Percent, Fractions, and Decimals

✎ *Converting fractions to decimals.*

1) $\dfrac{50}{100}$

2) $\dfrac{38}{100}$

3) $\dfrac{15}{100}$

4) $\dfrac{80}{100}$

5) $\dfrac{7}{100}$

6) $\dfrac{35}{100}$

7) $\dfrac{90}{100}$

8) $\dfrac{20}{100}$

9) $\dfrac{7}{100}$

✎ *Write each decimal as a percent.*

10) 0.5

11) 0.9

12) 0.002

13) 0.524

14) 0.1

15) 0.03

16) 3.63

17) 0.008

18) 4.78

Percent Problems

✍ *Solve each problem.*

1) 51 is 340% of what?

2) 93% of what number is 97?

3) 27% of 142 is what number?

4) What percent of 125 is 29.3?

5) 60 is what percent of 126?

6) 67 is 67% of what?

7) 67 is 13% of what?

8) 41% of 78 is what?

9) 1 is what percent of 52.6?

10) What is 59% of 14 m?

11) What is 90% of 130 inches?

12) 16 inches is 35% of what?

13) 90% of 54.4 hours is what?

14) What percent of 33.5 is 21?

15) Liam scored 22 out of 30 marks in Algebra, 35 out of 40 marks in science and 89 out of 100 marks in mathematics. In which subject his percentage of marks in best?

16) Ella require 50% to pass. If she gets 280 marks and falls short by 20 marks, what were the maximum marks she could have got?

Markup, Discount, and Tax

✎ *Find the selling price of each item.*

1) Cost of a pen: $1.95, markup: 70%, discount: 40%, tax: 5%

2) Cost of a puppy: $349.99, markup: 41%, discount: 23%

3) Cost of a shirt: $14.95, markup: 25%, discount: 45%

4) Cost of an oil change: $21.95, markup: 95%

5) Cost of computer: $1,850.00, markup: 75%

Answers of Worksheets – Chapter 4

Percentage Calculations

1) 12.5
2) 12
3) 10.2
4) 31.5
5) 0
6) 17.6
7) 5.2
8) 42.12

9) 40
10) 2
11) 16
12) 0
13) 14
14) 33
15) 8
16) 176

17) 70
18) 72
19) 67%
20) 70%
21) 58%
22) 50%

Converting Between Percent, Fractions, and Decimals

1) 0.5
2) 0.38
3) 0.15
4) 0.8
5) 0.07
6) 0.35

7) 0.9
8) 0.2
9) 0.07
10) 50%
11) 90%
12) 0.2%

13) 52.4%
14) 10%
15) 3%
16) 363%
17) 0.8%
18) 478%

Percent Problems

1) 15
2) 104.3
3) 38.34
4) 23.44%
5) 47.6%
6) 100

7) 515.4
8) 31.98
9) 1.9%
10) 8.3 m
11) 117 inches
12) 45.7 inches

13) 49 hours
14) 62.7%
15) Mathematics
16) 600

58

Markup, Discount, and Tax

1) $2.09

2) $379.98

3) $10.28

4) $36.22

5) $3,237.50

Chapter 5: Algebraic Expressions

Math Topics that you'll learn today:

- ✓ Expressions and Variables
- ✓ Simplifying Variable Expressions
- ✓ Simplifying Polynomial Expressions
- ✓ Translate Phrases into an Algebraic Statement
- ✓ The Distributive Property
- ✓ Evaluating One Variable
- ✓ Evaluating Two Variables
- ✓ Combining like Terms

Without mathematics, there's nothing you can do. Everything around you is mathematics. Everything around you is numbers." – Shakuntala Devi

Expressions and Variables

✍ *Simplify each expression.*

1) $x + 5x$,

   use $x = 5$

5) $(-6)(-2x - 4y)$,

   use $x = 1$, $y = 3$

2) $8(-3x + 9) + 6$,

   use $x = 6$

6) $8x + 2 + 4y$,

   use $x = 9$, $y = 2$

3) $10x - 2x + 6 - 5$,

   use $x = 5$

7) $(-6)(-8x - 9y)$,

   use $x = 5$, $y = 5$

4) $2x - 3x - 9$,

   use $x = 7$

8) $6x + 5y$,

   use $x = 7$, $y = 4$

✍ *Simplify each expression.*

9) $5(-4 + 2x)$

12) $(-8)(6x - 4) + 12$

10) $-3 - 5x - 6x + 9$

13) $9(7x + 4) + 6x$

11) $6x - 3x - 8 + 10$

14) $(-9)(-5x + 2)$

Simplifying Variable Expressions

✍ *Simplify each expression.*

1) $-2 - x^2 - 6x^2$

2) $3 + 10x^2 + 2$

3) $8x^2 + 6x + 7x^2$

4) $5x^2 - 12x^2 + 8x$

5) $2x^2 - 2x - x$

6) $(-6)(8x - 4)$

7) $4x + 6(2 - 5x)$

8) $10x + 8(10x - 6)$

9) $9(-2x - 6) - 5$

10) $3(x + 9)$

11) $7x + 3 - 3x$

12) $2.5x^2 \times (-8x)$

✍ *Simplify.*

13) $-2(4 - 6x) - 3x, x = 1$

14) $2x + 8x, x = 2$

15) $9 - 2x + 5x + 2, x = 5$

16) $5(3x + 7), x = 3$

17) $2(3 - 2x) - 4, x = 6$

18) $5x + 3x - 8, x = 3$

19) $x - 7x, x = 8$

20) $5(-2 - 9x), x = 4$

Simplifying Polynomial Expressions

✍️ *Simplify each polynomial.*

1) $4x^5 - 5x^6 + 15x^5 - 12x^6 + 3\,x^6$

2) $(-3x^5 + 12 - 4x) + (8x^4 + 5x + 5\,x^5)$

3) $10x^2 - 5x^4 + 14x^3 - 20x^4 + 15x^3 - 8x^4$

4) $-6x^2 + 5x^2 - 7x^3 + 12 + 22$

5) $12x^5 - 5x^3 + 8x^2 - 8x^5$

6) $5x^3 + 1 + x^2 - 2x - 10x$

7) $14x^2 - 6x^3 - 2x\,(4x^2 + 2x)$

8) $(4x^4 - 2x) - (4x - 2x^4)$

9) $(3x^2 + 1) - (4 + 2x^2)$

10) $(2x + 2) - (7x + 6)$

11) $(12x^3 + 4x^4) - (2x^4 - 6x^3)$

12) $(12 + 3x^3) + (6x^3 + 6)$

13) $(5x^2 - 3) + (2x^2 - 3x^3)$

14) $(23x^3 - 12x^2) - (2x^2 - 9x^3)$

15) $(4x - 3x^3) - (3x^3 + 4x)$

Translate Phrases into an Algebraic Statement

✏️ *Write an algebraic expression for each phrase.*

1) A number increased by forty–two.

2) The sum of fifteen and a number

3) The difference between fifty–six and a number.

4) The quotient of thirty and a number.

5) Twice a number decreased by 25.

6) Four times the sum of a number and − 12.

7) A number divided by − 20.

8) The quotient of 60 and the product of a number and − 5.

9) Ten subtracted from a number.

10) The difference of six and a number.

The Distributive Property

✏️*Use the distributive property to simply each expression.*

1) $-(-2-5x)$

2) $(-6x+2)(-1)$

3) $(-5)(x-2)$

4) $-(7-3x)$

5) $8(8+2x)$

6) $2(12+2x)$

7) $(-6x+8)4$

8) $(3-6x)(-7)$

9) $(-12)(2x+1)$

10) $(8-2x)9$

11) $(-2x)(-1+9x)-4x(4+5x)$

12) $3(-5x-3)+4(6-3x)$

13) $(-2)(x+4)-(2+3x)$

14) $(-4)(3x-2)+6(x+1)$

15) $(-5)(4x-1)+4(x+2)$

16) $(-3)(x+4)-(2+3x)$

Evaluating One Variable

🖎 *Simplify each algebraic expression.*

1) $9 - x$, $x = 3$

2) $x + 2$, $x = 5$

3) $3x + 7$, $x = 6$

4) $x + (-5)$, $x = -2$

5) $3x + 6$, $x = 4$

6) $4x + 6$, $x = -1$

7) $10 + 2x - 6$, $x = 3$

8) $10 - 3x$, $x = 8$

9) $\frac{20}{x} - 3$, $x = 5$

10) $(-3) + \frac{x}{4} + 2x$, $x = 16$

11) $(-2) + \frac{x}{7}$, $x = 21$

12) $(-\frac{14}{x}) - 9 + 4x$, $x = 2$

13) $(-\frac{6}{x}) - 9 + 2x$, $x = 3$

14) $(-2) + \frac{x}{8}$, $x = 16$

15) $8(5x - 12)$, $x = -2$

Evaluating Two Variables

✎*Simplify each algebraic expression.*

1) $2x + 4y - 3 + 2$,

   $x = 5, y = 3$

6) $6 + 3(-2x - 3y)$,

   $x = 9, y = 7$

2) $(-\frac{12}{x}) + 1 + 5y$,

   $x = 6, y = 8$

7) $12x + y$,

   $x = 4, y = 8$

3) $(-4)(-2a - 2b)$,

   $a = 5, b = 3$

8) $x \times 4 \div y$,

   $x = 3, y = 2$

4) $10 + 3x + 7 - 2y$,

   $x = 7, y = 6$

9) $2x + 14 + 4y$,

   $x = 6, y = 8$

5) $9x + 2 - 4y$,

   $x = 7, y = 5$

10) $4a - (5 - b)$,

    $a = 4, b = 6$

Combining like Terms

✍ *Simplify each expression.*

1) $5 + 2x - 8$

2) $(-2x + 6)\,2$

3) $7 + 3x + 6x - 4$

4) $(-4) - (3)(5x + 8)$

5) $9x - 7x - 5$

6) $x - 12x$

7) $7\,(3x + 6) + 2x$

8) $(-11x) - 10x$

9) $3x - 12 - 5x$

10) $13 + 4x - 5$

11) $(-22x) + 8x$

12) $2\,(4 + 3x) - 7x$

13) $(-4x) - (6 - 14x)$

14) $5\,(6x - 1) + 12x$

15) $22x + 6 + 2x$

16) $(-13x) - 14x$

17) $(-6x) - 9 + 15x$

18) $(-6x) + 7x$

19) $(-5x) + 12 + 7x$

20) $(-3x) - 9 + 15x$

21) $20x - 19x$

Answers of Worksheets – Chapter 5

Expressions and Variables

1) 30
2) −66
3) 41
4) −16
5) 84

6) 82
7) 510
8) 62
9) 10x − 20
10) 6 − 11x

11) 3x + 2
12) 44 − 48x
13) 69x + 36
14) 45x − 18

Simplifying Variable Expressions

1) $-7x^2 - 2$
2) $10x^2 + 5$
3) $15x^2 + 6x$
4) $-7x^2 + 8x$
5) $2x^2 - 3x$
6) $-48x + 24$
7) $-26x + 12$

8) $90x - 48$
9) $-18x - 59$
10) $3x + 27$
11) $4x + 3$
12) $-20x^3$
13) 1
14) 20

15) 26
16) 80
17) -22
18) 16
19) -48
20) -190

Simplifying Polynomial Expressions

1) $-14x^6 + 19x^5$
2) $2x^5 + 8x^4 + x + 12$
3) $-33x^4 + 29x^3 + 10x^2$
4) $-7x^3 - x^2 + 34$
5) $4x^5 - 5x^3 + 8x^2$
6) $5x^3 + x^2 - 12x + 1$
7) $-14x^3 + 10x^2$
8) $6x^4 - 6x$

9) $x^2 - 3$
10) $-5x - 4$
11) $2x^4 + 18x^3$
12) $9x^3 + 18$
13) $-3x^3 + 7x^2 - 3$
14) $32x^3 - 14x^2$
15) $-6x^3$

Translate Phrases into an Algebraic Statement

1) $x + 42$
3) $56 - x$
4) $30/x$
5) $2x - 25$
8) $\dfrac{60}{-5x}$

2) $15 + x$
6) $4(x + (-12))$
7) $\dfrac{x}{-20}$
9) $x - 10$
10) $6 - x$

The Distributive Property

1) $5x + 2$
2) $6x - 2$
3) $-5x + 10$
4) $3x - 7$
5) $16x + 64$
6) $4x + 24$
7) $- 24x + 32$
8) $42x - 21$
9) $- 24x - 12$
10) $- 18x + 72$
11) $- 38x^2 - 14x$
12) $- 27x + 15$
13) $- 5x - 10$
14) $- 6x + 14$
15) $- 16x + 13$
16) $- 6x - 14$

Evaluating One Variable

1) 6
2) 7
3) 25
4) -7
5) 18
6) 2
7) 10
8) -14
9) 1
10) 33
11) 1
12) -8
13) -5
14) 0
15) -176

Evaluating Two Variables

1) 21
2) 39
3) 64
4) 26
5) 45
6) -111
7) 56
8) 6
9) 58
10) 17

Combining like Terms

1) $2x - 3$
2) $-4x + 12$
3) $9x + 3$
4) $-15x - 28$
5) $2x - 5$
6) $-11x$
7) $23x + 42$
8) $-21x$
9) $-2x - 12$
10) $4x + 8$
11) $-14x$
12) $- x + 8$
13) $10x - 6$
14) $42x - 5$
15) $24x + 6$
16) $-27x$
17) $9x - 9$
18) x
19) $2x + 12$
20) $12x - 9$
21) x

Chapter 6: Equations

Math Topics that you'll learn today:

- ✓ One– Step Equations
- ✓ Two– Step Equations
- ✓ Multi– Step Equations

"The study of mathematics, like the Nile, begins in minuteness but ends in magnificence."

– Charles Caleb Colton

One–Step Equations

✍ *Solve each equation.*

1) $x + 3 = 17$

2) $22 = (-8) + x$

3) $3x = (-30)$

4) $(-36) = (-6x)$

5) $(-6) = 4 + x$

6) $2 + x = (-2)$

7) $20x = (-220)$

8) $18 = x + 5$

9) $(-23) + x = (-19)$

10) $5x = (-45)$

11) $x - 12 = (-25)$

12) $x - 3 = (-12)$

13) $(-35) = x - 27$

14) $8 = 2x$

15) $(-6x) = 36$

16) $(-55) = (-5x)$

17) $x - 30 = 20$

18) $8x = 32$

19) $36 = (-4x)$

20) $4x = 68$

21) $30x = 300$

Two–Step Equations

✍ *Solve each equation.*

1) $5(8 + x) = 20$

2) $(-7)(x - 9) = 42$

3) $(-12)(2x - 3) = (-12)$

4) $6(1 + x) = 12$

5) $12(2x + 4) = 60$

6) $7(3x + 2) = 42$

7) $8(14 + 2x) = (-34)$

8) $(-15)(2x - 4) = 48$

9) $3(x + 5) = 12$

10) $\dfrac{3x - 12}{6} = 4$

11) $(-12) = \dfrac{x + 15}{6}$

12) $110 = (-5)(2x - 6)$

13) $\dfrac{x}{8} - 12 = 4$

14) $20 = 12 + \dfrac{x}{4}$

15) $\dfrac{-24 + x}{6} = (-12)$

16) $(-4)(5 + 2x) = (-100)$

17) $(-12x) + 20 = 32$

18) $\dfrac{-2 + 6x}{4} = (-8)$

19) $\dfrac{x + 6}{5} = (-5)$

20) $(-9) + \dfrac{x}{4} = (-15)$

Multi–Step Equations

✎ *Solve each equation.*

1) $-(2 - 2x) = 10$

2) $-12 = -(2x + 8)$

3) $3x + 15 = (-2x) + 5$

4) $-28 = (-2x) - 12x$

5) $2(1 + 2x) + 2x = -118$

6) $3x - 18 = 22 + x - 3 + x$

7) $12 - 2x = (-32) - x + x$

8) $7 - 3x - 3x = 3 - 3x$

9) $6 + 10x + 3x = (-30) + 4x$

10) $(-3x) - 8(-1 + 5x) = 352$

11) $24 = (-4x) - 8 + 8$

12) $9 = 2x - 7 + 6x$

13) $6(1 + 6x) = 294$

14) $-10 = (-4x) - 6x$

15) $4x - 2 = (-7) + 5x$

16) $5x - 14 = 8x + 4$

17) $40 = -(4x - 8)$

18) $(-18) - 6x = 6(1 + 3x)$

19) $x - 5 = -2(6 + 3x)$

20) $6 = 1 - 2x + 5$

Answers of Worksheets – Chapter 6

One–Step Equations

1) 14	8) 13	15) − 6
2) 30	9) 4	16) 11
3) − 10	10) − 9	17) 50
4) 6	11) − 13	18) 4
5) − 10	12) − 9	19) − 9
6) − 4	13) − 8	20) 17
7) − 11	14) 4	21) 10

Two–Step Equations

1) − 4	8) $\frac{2}{5}$	15) − 48
2) 3	9) − 1	16) 10
3) 2	10) 12	17) − 1
4) 1	11) − 87	18) − 5
5) 0.5	12) − 8	19) − 31
6) $\frac{4}{3}$	13) 128	20) − 24
7) $-\frac{73}{8}$	14) 32	

Multi–Step Equations

1) 6	8) $\frac{4}{3}$	14) 1
2) 2	9) − 4	15) 5
3) − 2	10) − 8	16) − 6
4) 2	11) − 6	17) − 8
5) − 20	12) 2	18) − 1
6) 37	13) 8	19) − 1
7) 22		20) 0

Chapter 7: Inequalities

Math Topics that you'll learn today:

- ✓ Graphing Single– Variable Inequalities
- ✓ One– Step Inequalities
- ✓ Two– Step Inequalities
- ✓ Multi– Step Inequalities

Without mathematics, there's nothing you can do. Everything around you is mathematics.
Everything around you is numbers." – Shakuntala Devi

Graphing Single–Variable Inequalities

✍ *Draw a graph for each inequality.*

1) $-2 > x$

2) $5 \leq -x$

3) $x > 7$

4) $-x > 1.5$

One–Step Inequalities

✍ *Solve each inequality and graph it.*

1) $x + 9 \geq 11$

2) $x - 4 \leq 2$

3) $6x \geq 36$

4) $7 + x < 16$

5) $x + 8 \leq 1$

6) $3x > 12$

7) $3x < 24$

Two–Step Inequalities

✍️*Solve each inequality and graph it.*

1) $3x - 4 \leq 5$

2) $2x - 2 \leq 6$

3) $4x - 4 \leq 8$

4) $3x + 6 \geq 12$

5) $6x - 5 \geq 19$

6) $2x - 4 \leq 6$

7) $8x - 4 \leq 4$

8) $6x + 4 \leq 10$

9) $5x + 4 \leq 9$

10) $7x - 4 \leq 3$

11) $4x - 19 < 19$

12) $2x - 3 < 21$

13) $7 + 4x \geq 19$

14) $9 + 4x < 21$

15) $3 + 2x \geq 19$

16) $6 + 4x < 22$

Multi–Step Inequalities

✍️*Solve each inequality.*

1) $\dfrac{9x}{7} - 7 < 2$

2) $\dfrac{4x + 8}{2} \le 12$

3) $\dfrac{3x - 8}{7} > 1$

4) $-3(x - 7) > 21$

5) $4 + \dfrac{x}{3} < 7$

6) $\dfrac{2x + 6}{4} \le 10$

Answers of Worksheets – Chapter 7

Graphing Single–Variable Inequalities

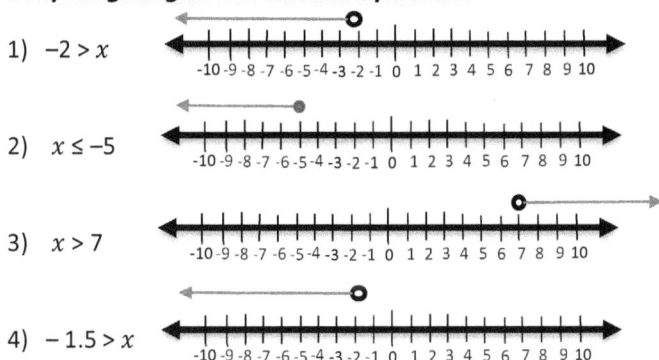

1) $-2 > x$

2) $x \leq -5$

3) $x > 7$

4) $-1.5 > x$

One–Step Inequalities

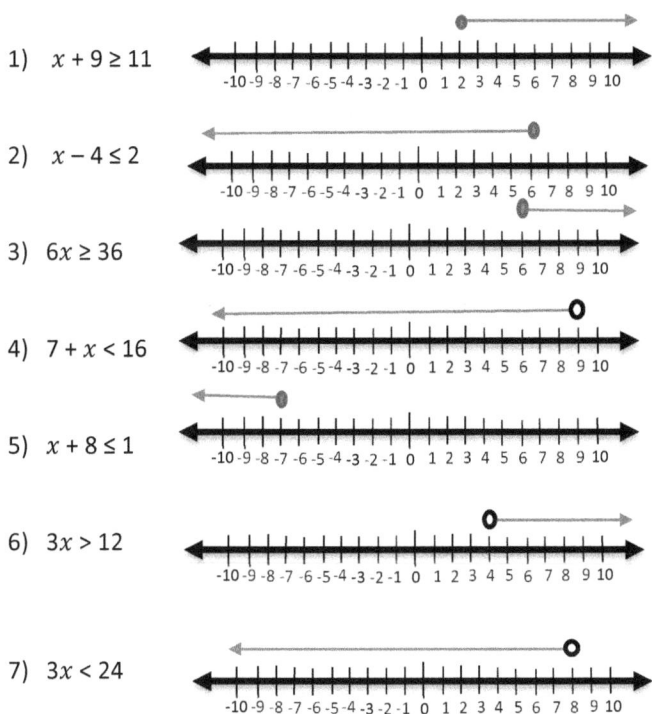

1) $x + 9 \geq 11$

2) $x - 4 \leq 2$

3) $6x \geq 36$

4) $7 + x < 16$

5) $x + 8 \leq 1$

6) $3x > 12$

7) $3x < 24$

Two–Step inequalities

1) $x \leq 3$

2) $x \leq 4$

3) $x \leq 3$

4) $x \geq 2$

5) $x \geq 4$

6) $x \leq 5$

7) $x \leq 1$

8) $x \leq 1$

9) $x \leq 1$

10) $x \leq 1$

11) $x < 9.5$

12) $x < 12$

13) $x \geq 3$

14) $x < 3$

15) $x \geq 8$

16) $x < 4$

Multi–Step inequalities

1) $x < 7$

2) $x \leq 4$

3) $x > 5$

4) $x < 0$

5) $x < 9$

6) $x \leq 17$

Chapter 8: Linear

Functions

Math Topics that you'll learn today:

- ✓ Finding Slope
- ✓ Graphing Lines Using Slope– Intercept Form
- ✓ Graphing Lines Using Standard Form
- ✓ Writing Linear Equations
- ✓ Graphing Linear Inequalities
- ✓ Finding Midpoint
- ✓ Finding Distance of Two Points

"Sometimes the questions are complicated and the answers are simple." – Dr. Seuss

inding Slope

✍ *Find the slope of the line through each pair of points.*

1) $(1, 1), (3, 5)$

2) $(4, -6), (-3, -8)$

3) $(7, -12), (5, 10)$

4) $(19, 3), (20, 3)$

5) $(15, 8), (-17, 9)$

6) $(6, -12), (15, -3)$

7) $(3, 1), (7, -5)$

8) $(3, -2), (-7, 8)$

9) $(15, -3), (-9, 5)$

10) $(-4, 7), (-6, -4)$

11) $(6, -8), (-11, -7)$

12) $(-6, 13), (17, -9)$

13) $(-10, -2), (-6, -5)$

14) $(4, 5), (-4, 10)$

15) $(-3, 1), (-17, 2)$

16) $(7, 0), (-13, -11)$

17) $(17, -13), (17, 8)$

18) $(12, 2), (-7, 5)$

Graphing Lines Using Slope–Intercept Form

Example:

$y = 8x - 3$

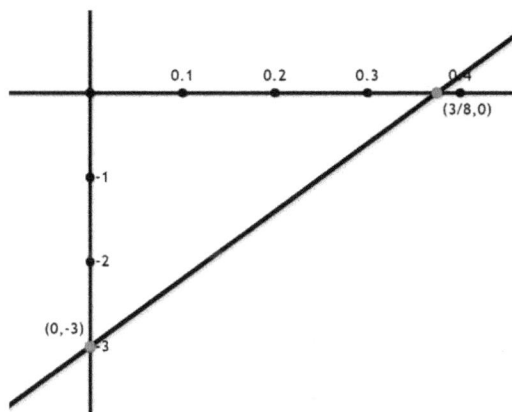

✍ *Sketch the graph of each line.*

1) $y = \dfrac{1}{2}x - 4$

2) $y = 2x$

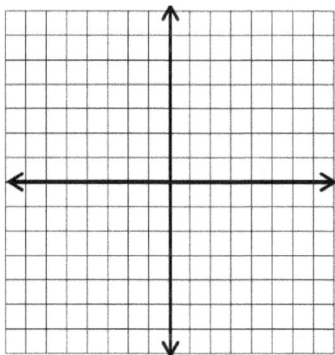

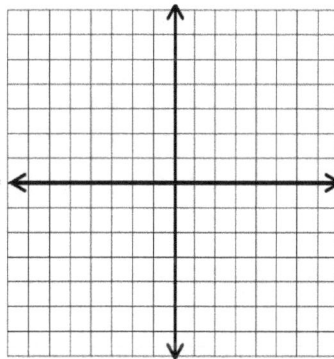

Graphing Lines Using Standard Form

Example:

$x + 4y = 12$

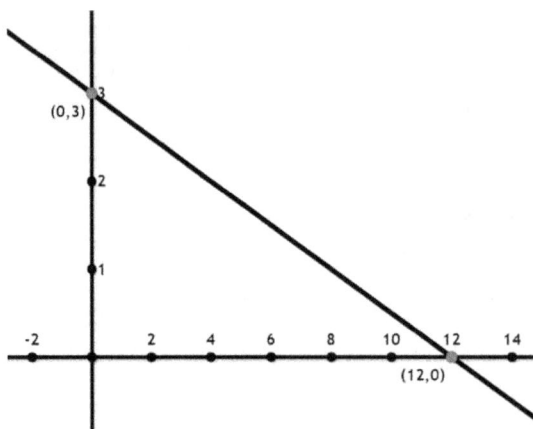

✍ *Sketch the graph of each line.*

1) $2x - y = 4$

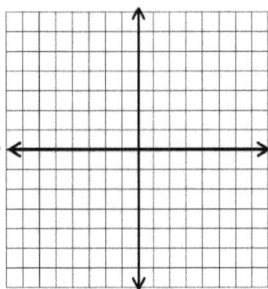

2) $x + y = 2$

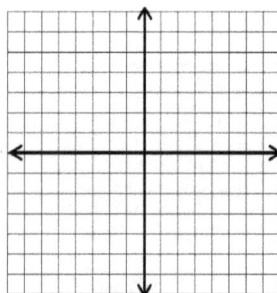

Writing Linear Equations

✍ *Write the slope–intercept form of the equation of the line through the given points.*

1) through: $(-4, -2), (-3, 5)$

2) through: $(5, 4), (-4, 3)$

3) through: $(0, -2), (-5, 3)$

4) through: $(-1, 1), (-2, 6)$

5) through: $(0, 3), (-4, -1)$

6) through: $(0, 2), (1, -3)$

7) through: $(0, -5), (4, 3)$

8) through: $(-1, 4), (0, 4)$

9) through: $(2, -3), (3, -5)$

10) through: $(2, 5), (-1, -4)$

11) through: $(1, -3), (-3, 1)$

12) through: $(3, 3), (1, -5)$

13) through: $(4, 4), (3, -5)$

14) through: $(0, 3), (1, 1)$

15) through: $(5, 5), (2, -3)$

16) through: $(-2, -2), (2, -5)$

17) through: $(-3, -2), (1, -1)$

18) through: $(-2, 1), (6, 5)$

Graphing Linear Inequalities

✍️*Sketch the graph of each linear inequality.*

1) $y < -4x + 2$

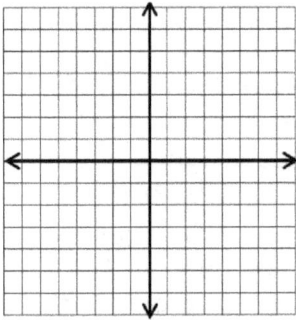

2) $2x + y < -4$

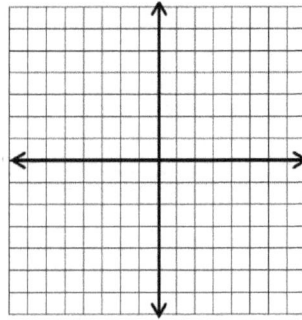

4) $x - 3y < -5$

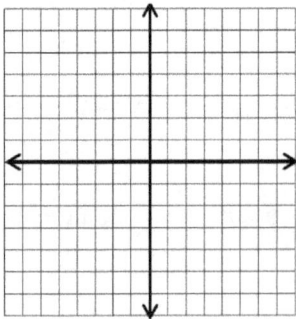

5) $6x - 2y \geq 8$

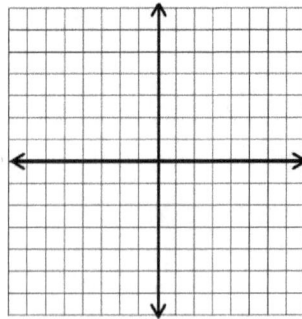

Finding Midpoint

✎*Find the midpoint of the line segment with the given endpoints.*

1) $(2, -2), (3, -5)$

2) $(0, 2), (-2, -6)$

3) $(7, 4), (9, -1)$

4) $(4, -5), (0, 8)$

5) $(1, -2), (1, -6)$

6) $(-2, -3), (3, -6)$

7) $(7, 0), (-7, 5)$

8) $(-2, 6), (-3, -2)$

9) $(-1, 1), (5, -5)$

10) $(2.3, -1.3), (-2.2, -0.5)$

11) $(4.1, 6.32), (4, 5.6)$

12) $(2, -1), (-6, 0)$

13) $(-4, 4), (5, -1)$

14) $(-2, -3), (-6, 5)$

15) $(\frac{1}{2}, 1), (2, 4)$

16) $(-2, -2), (6, 5)$

Finding Distance of Two Points

✎ *Find the distance between each pair of points.*

1) $(2, -1), (1, -1)$

2) $(6, 4), (-1, 3)$

3) $(-8, -5), (-6, 1)$

4) $(-6, -10), (-2, -10)$

5) $(4, -6), (-3, 4)$

6) $(-6, -7), (-2, -8)$

7) $(5, 4), (8, 2)$

8) $(8, 4), (3, -7)$

9) $(1, 3), (5, 7)$

10) $(4, 2), (-7, 1)$

11) $(-3, -4), (-7, -2)$

12) $(-7, -2), (6, 9)$

13) $(10, 0), (0, 4)$

14) $(-3, 2), (5, 0)$

15) $(-5, 6), (8, -4)$

16) $(3, -5), (-8, -4)$

17) $(0, 8), (4, 10)$

18) $(6, 4), (-5, -1)$

Answers of Worksheets – Chapter 8

Finding Slope

1) 2

2) $\frac{2}{7}$

3) −11

4) 0

5) $-\frac{1}{32}$

6) 1

7) $-\frac{3}{2}$

8) −1

9) $-\frac{1}{3}$

10) $\frac{11}{2}$

11) $-\frac{1}{17}$

12) $-\frac{22}{23}$

13) $-\frac{3}{4}$

14) $-\frac{5}{8}$

15) $-\frac{1}{14}$

16) $\frac{11}{20}$

17) Undefined

18) $-\frac{3}{19}$

Graphing Lines Using Slope–Intercept Form

1)

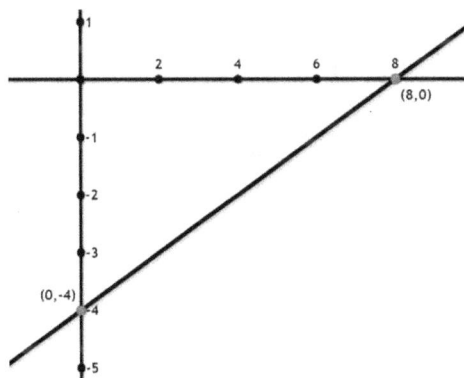

2)

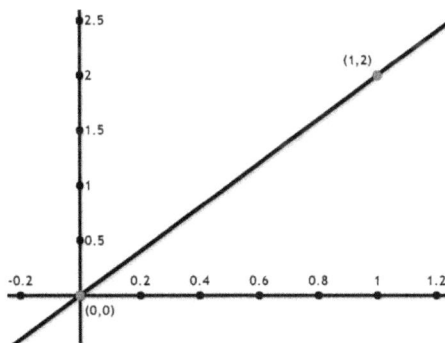

Graphing Lines Using Standard Form

1)

2)

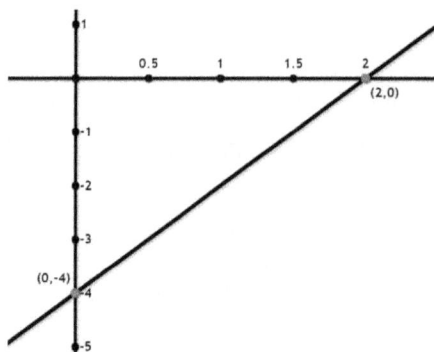
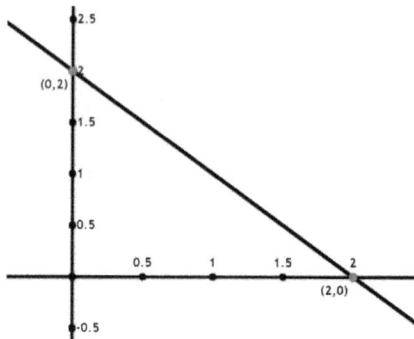

Writing Linear Equations

1) $y = 7x + 26$

2) $y = \frac{1}{9}x + \frac{31}{9}$

3) $y = -x - 2$

4) $y = -5x - 4$

5) $y = x + 3$

6) $y = -5x + 2$

7) $y = 2x - 5$

8) $y = 4$

9) $y = -2x + 1$

10) $y = 3x - 1$

11) $y = -x - 2$

12) $y = 4x - 9$

13) $y = 9x - 32$

14) $y = -2x + 3$

15) $y = \frac{8}{3}x - \frac{25}{3}$

16) $y = -\frac{3}{4}x - \frac{7}{2}$

17) $y = \frac{1}{4}x - \frac{5}{4}$

18) $y = -\frac{4}{3}x + \frac{19}{3}$

Graphing Linear Inequalities

1)

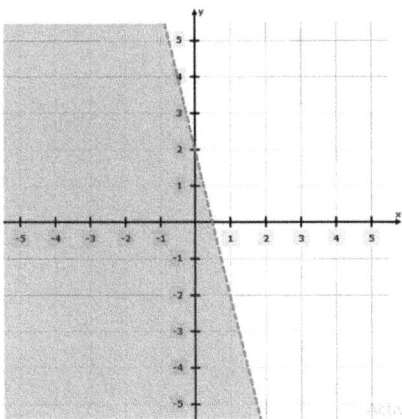

2)

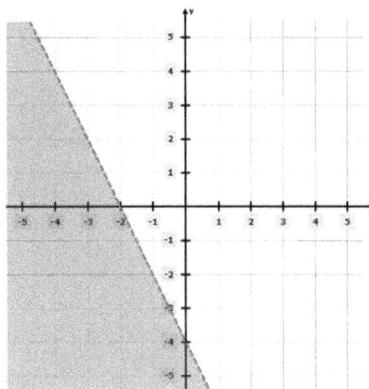

4)

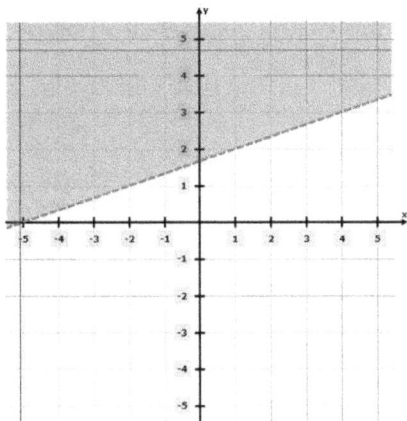

5)

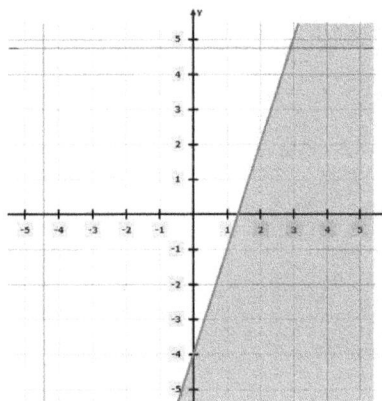

Finding Midpoint

1) $(2.5, -3.5)$

2) $(-1, -2)$

3) $(8, 1.5)$

4) $(2, 1.5)$

5) $(1, -4)$

6) $(0.5, -4.5)$

7) $(0, 2.5)$

8) $(-2.5, 2)$

9) $(2, -2)$

10) $(0.05, -0.9)$

11) $(4.05, 5.96)$

12) $(-2, -0.5)$

13) $(\frac{1}{2}, 1\frac{1}{2})$

14) $(-4, 1)$

15) $(1.25, 2.5)$

16) $(2, \frac{3}{2})$

Finding Distance of Two Points

1) 1

2) 7.1

3) 6.32

4) 4

5) 12.21

6) 4.12

7) 3.61

8) 12.1

9) 5.66

10) 11.04

11) 4.47

12) 17.03

13) 10.77

14) 8.25

15) 16.4

16) 10.3

17) 4.47

18) 12.1

Chapter 9: Polynomials

Math Topics that you'll learn today:

- ✓ Classifying Polynomials
- ✓ Writing Polynomials in Standard Form
- ✓ Simplifying Polynomials
- ✓ Adding and Subtracting Polynomials
- ✓ Multiplying Monomials
- ✓ Multiplying and Dividing Monomials
- ✓ Multiplying a Polynomial and a Monomial
- ✓ Multiplying Binomials
- ✓ Factoring Trinomials
- ✓ Operations with Polynomials

Mathematics – the unshaken Foundation of Sciences, and the plentiful Fountain of Advantage to human

affairs. — Isaac Barrow

Classifying Polynomials

✎ *Name each polynomial by degree and number of terms.*

1) x

2) $-5x^4$

3) $7x - 4$

4) -6

5) $8x + 1$

6) $9x^2 - 8x^3$

7) $2x^5$

8) $10 + 8x$

9) $5x^2 - 6x$

10) $-7x^7 + 7x^4$

11) $-8x^4 + 5x^3 - 2x^2 - 8x$

12) $4x - 9x^2 + 4x^3 - 5x^4$

13) $4x^6 + 5x^5 + x^4$

14) $-4 - 2x^2 + 8x$

15) $9x^6 - 8$

16) $7x^5 + 10x^4 - 3x + 10x^7$

17) $4x^6 - 3x^2 - 8x^4$

18) $-5x^4 + 10x - 10$

Writing Polynomials in Standard Form

✎ *Write each polynomial in standard form.*

1) $3x^2 - 5x^3$

2) $3 + 4x^3 - 3$

3) $2x^2 + 1x - 6x^3$

4) $9x - 7x$

5) $12 - 7x + 9x^4$

6) $5x^2 + 13x - 2x^3$

7) $-3 + 16x - 16x$

8) $3x \, (x + 4) - 2 \, (x + 4)$

9) $(x + 5) \, (x - 2)$

10) $3x^2 + x + 12 - 5x^2 - 2x$

11) $12x^5 + 7x^3 - 3x^5 - 8x^3$

12) $3x \, (2x + 5 - 2x^2)$

13) $11x \, (x^5 + 2x^3)$

14) $(x + 6) \, (x + 3)$

15) $(x + 4)^2$

16) $(8x - 7) \, (3x + 2)$

17) $5x \, (3x^2 + 2x + 1)$

18) $7x \, (3 - x + 6x^3)$

Simplifying Polynomials

✎ *Simplify each expression.*

1) $11 - 4x^2 + 3x^2 - 7x^3 + 3$

2) $2x^5 - x^3 + 8x^2 - 2x^5$

3) $(-5)(x^6 + 10) - 8(14 - x^6)$

4) $4(2x^2 + 4x^2 - 3x^3) + 6x^3 + 17$

5) $11 - 6x^2 + 5x^2 - 12x^3 + 22$

6) $2x^2 - 2x + 3x^3 + 12x - 22x$

7) $(3x - 8)(3x - 4)$

8) $(12x + 2y)^2$

9) $(12x^3 + 28x^2 + 10x + 4) \div (x + 2)$

10) $(2x + 12x^2 - 2) \div (2x + 1)$

11) $(2x^3 - 1) + (3x^3 - 2x^3)$

12) $(x - 5)(x - 3)$

13) $(3x + 8)(3x - 8)$

14) $(8x^2 - 3x) - (5x - 5 - 8x^2)$

Adding and Subtracting Polynomials

✎ *Simplify each expression.*

1) $(2x^3 - 2) + (2x^3 + 2)$

2) $(4x^3 + 5) - (7 - 2x^3)$

3) $(4x^2 + 2x^3) - (2x^3 + 5)$

4) $(4x^2 - x) + (3x - 5x^2)$

5) $(7x + 9) - (3x + 9)$

6) $(4x^4 - 2x) - (6x - 2x^4)$

7) $(12x - 4x^3) - (8x^3 + 6x)$

8) $(2x^3 - 8x^2) - (5x^2 - 3x^3)$

9) $(2x^2 - 6) + (9x^2 - 4x^3)$

10) $(4x^3 + 3x^4) - (x^4 - 5x^3)$

11) $(-12x^4 + 10x^5 + 2x^3) + (14x^3 + 23x^5 + 8x^4)$

12) $(13x^2 - 6x^5 - 2x) - (-10x^2 - 11x^5 + 9x)$

13) $(35 + 9x^5 - 3x^2) + (8x^4 + 3x^5) - (27 - 5x^4)$

14) $(3x^5 - 2x^3 - 4x) + (4x + 10x^4 - 23) + (x^2 - x^3 + 12)$

Multiplying Monomials

✍ *Simplify each expression.*

1) $2xy^2z \times 4z^2$

2) $4xy \times x^2y$

3) $4pq^3 \times (-2p^4q)$

4) $8s^4t^2 \times st^5$

5) $12p^3 \times (-3p^4)$

6) $-4p^2q^3r \times 6pq^2r^3$

7) $(-8a^4) \times (-12a^6b)$

8) $3u^4v^2 \times (-7u^2v^3)$

9) $4u^3 \times (-2u)$

10) $-6xy^2 \times 3x^2y$

11) $12y^2z^3 \times (-y^2z)$

12) $5a^2bc^2 \times 2abc^2$

Multiplying and Dividing Monomials

Example:

$(-3x^2)(8x^4y^{12}) = -24x^6y^{12}$

$\dfrac{36\,x^5y^7}{4\,x^4y^5} = 9xy^2$

✍ *Simplify.*

1) $(7x^4y^6)(4x^3y^4)$

2) $(15x^4)\,(3x^9)$

3) $(12x^2y^9)(7x^9y^{12})$

4) $\dfrac{80\,x^{12}y^9}{10\,x^6y^7}$

5) $\dfrac{95\,x^{18}y^7}{5\,x^9y^2}$

6) $\dfrac{200\,x^3y^8}{40\,x^3y^7}$

7) $\dfrac{-15\,x^{17}y^{13}}{3\,x^6y^9}$

8) $\dfrac{-64\,x^8y^{10}}{8\,x^3y^7}$

Multiplying a Polynomial and a Monomial

✎ *Find each product.*

1) $5(3x - 6y)$

2) $9x(2x + 4y)$

3) $8x(7x - 4)$

4) $12x(3x + 9)$

5) $11x(2x - 11y)$

6) $2x(6x - 6y)$

7) $3x(2x^2 - 3x + 8)$

8) $13x(4x + 8y)$

9) $20(2x^2 - 8x - 5)$

10) $3x(3x - 2)$

11) $6x^3(3x^2 - 2x + 2)$

12) $8x^2(3x^2 - 5xy + 7y^2)$

13) $2x^2(3x^2 - 5x + 12)$

14) $2x^3(2x^2 + 5x - 4)$

15) $5x(6x^2 - 5xy + 2y^2)$

16) $9(x^2 + xy - 8y^2)$

Multiplying Binomials

✍ *Multiply.*

1) $(3x - 2)(4x + 2)$

2) $(2x - 5)(x + 7)$

3) $(x + 2)(x + 8)$

4) $(x^2 + 2)(x^2 - 2)$

5) $(x - 2)(x + 4)$

6) $(x - 8)(2x + 8)$

7) $(5x - 4)(3x + 3)$

8) $(x - 7)(x - 6)$

9) $(6x + 9)(4x + 9)$

10) $(2x - 6)(5x + 6)$

11) $(x - 7)(x + 7)$

12) $(x + 4)(4x - 8)$

13) $(6x - 4)(6x + 4)$

14) $(x - 7)(x + 2)$

15) $(x - 8)(x + 8)$

16) $(3x + 3)(3x - 4)$

17) $(x + 3)(x + 3)$

18) $(x + 4)(x + 6)$

Factoring Trinomials

✍ *Factor each trinomial.*

1) $x^2 - 7x + 12$

2) $x^2 + 5x - 14$

3) $x^2 - 11x - 42$

4) $6x^2 + x - 12$

5) $x^2 - 17x + 30$

6) $x^2 + 8x + 15$

7) $3x^2 + 11x - 4$

8) $x^2 - 6x - 27$

9) $10x^2 + 33x - 7$

10) $x^2 + 24x + 144$

11) $49x^2 + 28xy + 4y^2$

12) $16x^2 - 40x + 25$

13) $x^2 - 10x + 25$

14) $25x^2 - 20x + 4$

15) $x^3 + 6x^2y^2 + 9xy^3$

16) $9x^2 + 24x + 16$

17) $x^2 - 8x + 16$

18) $x^2 + 121 + 22x$

Operations with Polynomials

✎ *Find each product.*

1) $3x^2 (6x - 5)$

2) $5x^2 (7x - 2)$

3) $-3 (8x - 3)$

4) $6x^3 (-3x + 4)$

5) $9 (6x + 2)$

6) $8 (3x + 7)$

7) $5 (6x - 1)$

8) $-7x^4 (2x - 4)$

9) $8 (x^2 + 2x - 3)$

10) $4 (4x^2 - 2x + 1)$

11) $2 (3x^2 + 2x - 2)$

12) $8x (5x^2 + 3x + 8)$

13) $(9x + 1) (3x - 1)$

14) $(4x + 5) (6x - 5)$

15) $(7x + 3) (5x - 6)$

16) $(3x - 4) (3x + 8)$

Answers of Worksheets – Chapter 9

Classifying Polynomials

1) Linear monomial
2) Quartic monomial
3) Linear binomial
4) Constant monomial
5) Linear binomial
6) Cubic binomial
7) Quantic monomial
8) Linear binomial
9) Quadratic binomial
10) Seventh degree binomial
11) Quartic polynomial with four terms
12) Quartic polynomial with four terms
13) Sixth degree trinomial
14) Quadratic trinomial
15) Sixth degree binomial
16) Seventh degree polynomial with four terms
17) Sixth degree trinomial
18) Quartic trinomial

Writing Polynomials in Standard Form

1) $-5x^3 + 3x^2$
2) $4x^3$
3) $-6x^3 + 2x^2 + x$
4) $2x$
5) $9x^4 - 7x + 12$
6) $-2x^3 + 5x^2 + 13x$
7) -3
8) $3x^2 + 10x - 8$
9) $x^2 + 3x - 10$
10) $-2x^2 - x + 12$
11) $9x^5 - x^3$
12) $-6x^3 + 6x^2 + 15x$
13) $11x^6 + 22x^4$
14) $x^2 + 9x + 18$
15) $x^2 + 8x + 16$
16) $24x^2 - 5x - 14$
17) $15x^3 + 10x^2 + 5x$
18) $42x^4 - 7x^2 + 21x$

Simplifying Polynomials

1) $-7x^3 - x^2 + 14$
2) $-x^3 + 8x^2$
3) $3x^6 - 162$
4) $-6x^3 + 24x^2 + 17$
5) $-12x^3 - x^2 + 33$
6) $3x^3 + 2x^2 - 12x$

7) $9x^2 - 36x + 32$

8) $144x^2 + 48xy + 4y^2$

9) $12x^2 + 4x + 2$

10) $6x - 1$

11) $3x^3 - 1$

12) $x^2 - 8x + 15$

13) $9x^2 - 64$

14) $16x^2 - 8x + 5$

Adding and Subtracting Polynomials

1) $4x^3$

2) $6x^3 - 2$

3) $4x^2 - 5$

4) $-x^2 + 2x$

5) $4x$

6) $6x^4 - 8x$

7) $-12x^3 + 6x$

8) $5x^3 - 13x^2$

9) $-4x^3 + 11x^2 - 6$

10) $2x^4 + 9x^3$

11) $33x^5 - 4x^4 + 16x^3$

12) $5x^5 + 23x^2 - 11x$

13) $12x^5 + 13x^4 - 3x^2 + 8$

14) $3x^5 + 10x^4 - 3x^3 + x^2 - 11$

Multiplying Monomials

1) $8xy^2z^3$

2) $4x^3y^2$

3) $-8p^5q^4$

4) $8s^5t^7$

5) $-36p^7$

6) $-24p^3q^5r^4$

7) $96a^{10}b$

8) $-21u^6v^5$

9) $-8u^4$

10) $-18x^3y^3$

11) $-12y^4z^4$

12) $10a^3b^2c^4$

Multiplying and Dividing Monomials

1) $28x^7y^{10}$

2) $45x^{13}$

3) $84x^{11}y^{21}$

4) $8x^6y^2$

5) $19x^9y^5$

6) $5y$

7) $-5x^{11}y^4$

8) $-8x^5y^3$

Multiplying a Polynomial and a Monomial

1) $15x - 30y$
2) $18x^2 + 36xy$
3) $56x^2 - 32x$
4) $36x^2 + 108x$
5) $22x^2 - 121xy$
6) $12x^2 - 12xy$
7) $6x^3 - 9x^2 + 24x$
8) $52x^2 + 104xy$
9) $40x^2 - 160x - 100$
10) $9x^2 - 6x$
11) $18x^5 - 12x^4 + 12x^3$
12) $24x^4 - 40x^3y + 56y^2x^2$
13) $6x^4 - 10x^3 + 24x^2$
14) $4x^5 + 10x^4 - 8x^3$
15) $30x^3 - 25x^2y + 10xy^2$
16) $9x^2 + 9xy - 72y^2$

Multiplying Binomials

1) $12x^2 - 2x - 4$
2) $2x^2 + 9x - 35$
3) $x^2 + 10x + 16$
4) $x^4 - 4$
5) $x^2 + 2x - 8$
6) $2x^2 - 8x - 64$
7) $15x^2 + 3x - 12$
8) $x^2 - 13x + 42$
9) $24x^2 + 90x + 81$
10) $10x^2 - 18x - 36$
11) $x^2 - 49$
12) $4x^2 + 8x - 32$
13) $36x^2 - 16$
14) $x^2 - 5x - 14$
15) $x^2 - 64$
16) $9x^2 - 3x - 12$
17) $x^2 + 6x + 9$
18) $x^2 + 10x + 24$

Factoring Trinomials

1) $(x - 3)(x - 4)$
2) $(x - 2)(x + 7)$
3) $(x + 3)(x - 14)$
4) $(2x + 3)(3x - 4)$
5) $(x - 15)(x - 2)$
6) $(x + 3)(x + 5)$
7) $(3x - 1)(x + 4)$
8) $(x - 9)(x + 3)$
9) $(5x - 1)(2x + 7)$
10) $(x + 12)(x + 12)$
11) $(7x + 2y)(7x + 2y)$
12) $(4x - 5)(4x - 5)$
13) $(x - 5)(x - 5)$
14) $(5x - 2)(5x - 2)$
15) $x(x^2 + 6xy^2 + 9y^3)$
16) $(3x + 4)(3x + 4)$
17) $(x - 4)(x - 4)$
18) $(x + 11)(x + 11)$

Operations with Polynomials

1) $18x^3 - 15x^2$

2) $35x^3 - 10x^2$

3) $-24x + 9$

4) $-18x^4 + 24x^3$

5) $54x + 18$

6) $24x + 56$

7) $30x - 5$

8) $-14x^5 + 28x^4$

9) $8x^2 + 16x - 24$

10) $16x^2 - 8x + 4$

11) $6x^2 + 4x - 4$

12) $40x^3 + 24x^2 + 64x$

13) $27x^2 - 6x - 1$

14) $24x^2 + 10x - 25$

15) $35x^2 - 27x - 18$

16) $9x^2 + 12x - 32$

Chapter 10: Quadratic and System of Equations

Math Topics that you'll learn today:

- ✓ Solve a Quadratic Equation
- ✓ Solving Systems of Equations by Substitution
- ✓ Solving Systems of Equations by Elimination
- ✓ Systems of Equations Word Problems

Mathematics is the door and key to the sciences. — Roger Bacon

Solve a Quadratic Equation

✍ *Solve each equation.*

1) $(x + 2)(x - 4) = 0$

2) $(x + 5)(x + 8) = 0$

3) $(3x + 2)(x + 3) = 0$

4) $(4x + 7)(2x + 5) = 0$

5) $x^2 - 11x + 19 = -5$

6) $x^2 + 7x + 18 = 8$

7) $x^2 - 10x + 22 = -2$

8) $x^2 + 3x - 12 = 6$

9) $18x^2 + 45x - 27 = 0$

10) $90x^2 - 84x = -18$

11) $x^2 + 8x = -15$

Solving Systems of Equations by Substitution

✍ *Solve each system of equation by substitution.*

1) $-2x + 2y = 4$

$-2x + y = 3$

4) $2y = -6x + 10$

$10x - 8y = -6$

2) $-10x + 2y = -6$

$6x - 16y = 48$

5) $3x - 9y = -3$

$3y = 3x - 3$

3) $y = -8$

$16x - 12y = 72$

6) $-4x + 12y = 12$

$-14x + 16y = -10$

Solving Systems of Equations by Elimination

✎ *Solve each system of equation by elimination.*

1) $10x - 9y = -12$

   $-5x + 3y = 6$

2) $-3x - 4y = 5$

   $x - 2y = 5$

3) $5x - 14y = 22$

   $-6x + 7y = 3$

4) $10x - 14y = -4$

   $-10x - 20y = -30$

5) $32x + 14y = 52$

   $16x - 4y = -40$

6) $2x - 8y = -6$

   $8x + 2y = 10$

7) $-4x + 4y = -4$

   $4x + 2y = 10$

8) $4x + 6y = 10$

   $8x + 12y = -20$

Systems of Equations Word Problems

Example:

The difference of two numbers is 6. Their sum is 14. Find the numbers.

$x + y = 6$

$x + y = 14 \, (10, 4)$

1) A farmhouse shelters 10 animals, some are pigs and some are ducks. Altogether there are 36 legs. How many of each animal are there?

2) A class 0f 195 students went on a field trip. They took vehicles, some cars and some buses. Find the number of cars and the number of buses they took if each car holds 5 students and each bus hold 45 students.

3) The sum of the digits of a certain two–digit number is 7. Reversing its increasing the number by 9. What is the number?

4) A boat traveled 336 miles downstream and back. The trip downstream took 12 hours. The trip back took 14 hours. What is the speed of the boat in still water? What is the speed of the current?

Answers of Worksheets – Chapter 10

Solving Quadratic Equations

1) $x = -2, x = 4$

2) $x = -5, x = -8$

3) $x = -\frac{2}{3}, x = -3$

4) $x = -\frac{7}{4}, x = -\frac{5}{2}$

5) $x = 8, x = 3$

6) $x = -5, x = -2$

7) $x = 6, x = 4$

8) $x = -6, x = 3$

9) $x = \frac{1}{2}, x = -3$

10) $x = \frac{3}{5}, x = \frac{1}{3}$

11) $x = -5, x = -3$

Solving Systems of Equations by Substitution

1) $(4, 9)$

2) $(-1, 1)$

3) $(0, -3)$

4) $(-24, -8)$

5) $(1, 2)$

6) $(4, 3)$

7) $(3, 2)$

8) $(-5, 1)$

Solving Systems of Equations by Elimination

1) $(15, 27)$

2) $(1, -2)$

3) $(-4, -3)$

4) $(1, 1)$

5) $(-1, 6)$

6) $(1, 1)$

7) $(2, 1)$

8) No solution

9) $(3, 4)$

10) $(4, 2)$

Systems of Equations Word Problems

1) $(2, 8)$

2) $(3, 4)$

3) $(10, 4)$

4) 34

5) boat: 26 mph, current: 2 mph

Chapter 11: Quadratic Functions

Math Topics that you'll learn today:

- ✓ Graphing Quadratic Functions in Standard Form

- ✓ Graphing Quadratic Functions in Vertex Form

- ✓ Solving Quadratic Equations

- ✓ Use the Quadratic Formula and the Discriminant

- ✓ Operations with Complex Numbers

- ✓ Solve Quadratic Inequalities

It's fine to work on any problem, so long as it generates interesting mathematics along the way – even if you don't solve it at the end of the day." – Andrew Wiles

Graphing Quadratic Functions in Standard Form

✎ *Sketch the graph of each function.*

1) $y > 2x^2$

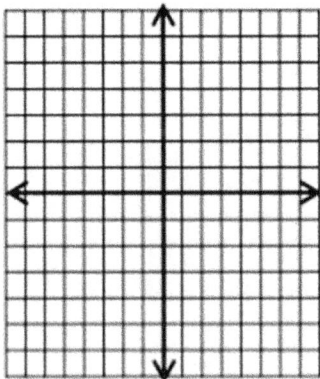

2) $y > 4x^2$

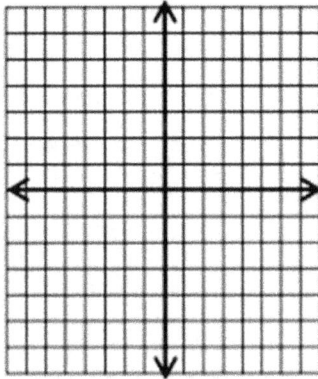

3) $2y < -4x^2$

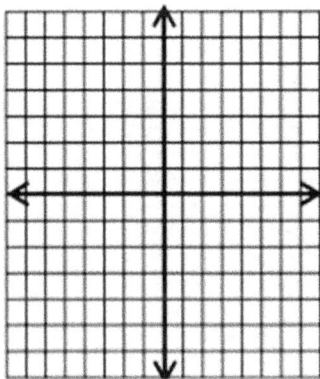

4) $y \geq 3x^2$

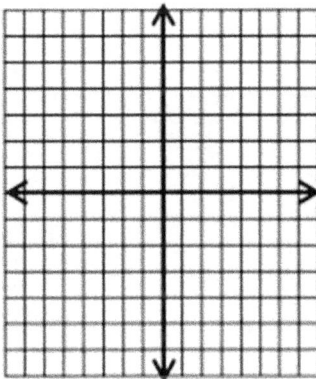

Graphing Quadratic Functions in Vertex Form

✍️ *Sketch the graph of each function. Identify the vertex and axis of symmetry.*

1) $y = 3(x + 1)^2 + 2$

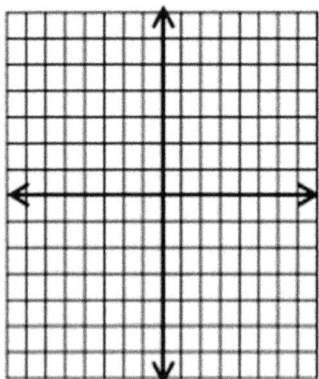

2) $y = -(x - 2)^2 - 4$

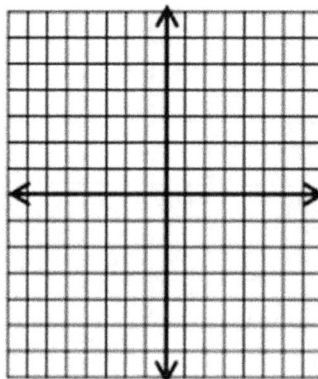

3) $y = 2(x - 3)^2 + 8$

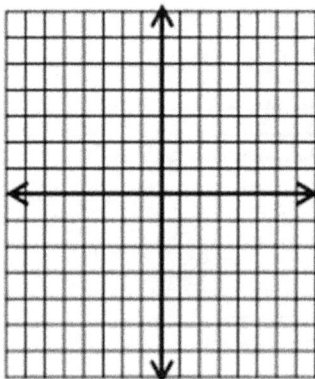

4) $y = x^2 - 8x + 19$

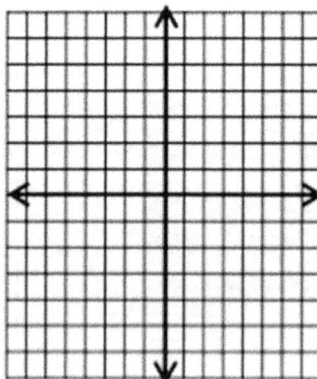

Solving Quadratic Equations by Factoring

✍ *Solve each equation by factoring.*

1) $x^2 + x - 20 = 2x$

2) $x^2 + 8x = -15$

3) $7x^2 - 14x = -7$

4) $6x^2 - 18x - 18 = 6$

5) $2x^2 + 6x - 24 = 12$

6) $2x^2 - 22x + 38 = -10$

7) $(2x + 5)(4x + 3) = 0$

8) $(x + 2)(x - 7) = 0$

9) $(x + 3)(x + 5) = 0$

10) $(5x + 7)(x + 4) = 0$

11) $-4x^2 - 8x - 3 = -3 - 5x^2$

12) $10x^2 = 27x - 18$

13) $7x^2 - 6x + 3 = 3$

14) $x^2 = 2x$

15) $2x^2 - 14 = -3x$

16) $10x^2 - 26x = -12$

17) $15x^2 + 80 = -80x$

18) $x^2 + 15x = -56$

Use the Quadratic Formula and the Discriminant

✎ *Find the value of the discriminant of each quadratic equation.*

1) $2x^2 + 5x - 4 = 0$

2) $x^2 + 5x + 2 = 0$

3) $5x^2 + x - 2 = 0$

4) $-4x^2 - 4x + 5 = 0$

5) $-2x^2 - x - 1 = 0$

6) $6x^2 - 2x - 3 = 0$

7) $x(x-1)$

8) $8x^2 - 9x = 0$

9) $3x^2 - 5x + 1 = 0$

10) $5x^2 + 6x + 4 = 0$

✎ *Find the discriminant of each quadratic equation then state the number of real and imaginary solution.*

11) $8x^2 - 6x + 3 = 5x^2$

12) $-4x^2 - 4x = 6$

13) $-x^2 - 9 = 6x$

14) $-9x^2 = -8x + 8$

15) $4x^2 = 8x - 4$

16) $9x^2 + 6x + 6 = 5$

17) $9x^2 - 3x - 8 = -10$

18) $-2x^2 - 8x - 14 = -6$

Operations with Complex Numbers

✎ *Simplify.*

1) $-3i \cdot 6i$

2) $8i \cdot i \cdot -2i$

3) $(5-3i)(3+i)$

4) $8i \cdot 2i(-5-3i)$

5) $(-2-i)(4+i)$

6) $(8-4i)(-9+5i)$

7) $(-5+3i)(-7-9i)$

8) $(8-6i)(-4-4i)$

9) $(5i)^3$

10) $6i + 8i \cdot i$

11) $-8(5-5i)$

12) $(8-3i)^2$

13) $6+4i-8i-8$

14) $-5i \cdot 2i - 5(-5+3i)$

15) $-4i(5-9i)(-2-8i)$

16) $8(-4+7i)(-4+5i)$

17) $(1-7i)^2$

18) $(2-4i)(-6+4i)$

Solve Quadratic Inequalities

✍ *Solve each quadratic inequality.*

1) $-x^2 - 5x + 6 > 0$

2) $x^2 - 5x - 6 < 0$

3) $x^2 + 4x - 5 > 0$

4) $x^2 - 2x - 3 \geq 0$

5) $x^2 - 1 < 0$

6) $17x^2 + 15x - 2 \geq 0$

7) $4x^2 + 20x - 11 < 0$

8) $12x^2 + 10x - 12 > 0$

9) $18x^2 + 23x + 5 \leq 0$

10) $-9x^2 + 29x - 6 \geq 0$

11) $-8x^2 + 6x - 1 \leq 0$

12) $5x^2 - 15x + 10 < 0$

13) $3x^2 - 5x \geq 4x^2 + 6$

14) $x^2 > 5x + 6$

15) $3x^2 + 7x \leq 5x^2 + 3x - 6$

16) $4x^2 - 12 > 3x^2 + x$

17) $3x^2 - 5x \geq 4x^2 + 6$

18) $2x^2 + 2x - 8 > x^2$

Answers of Worksheets – Chapter 11

Graphing quadratic functions in standard form

1)

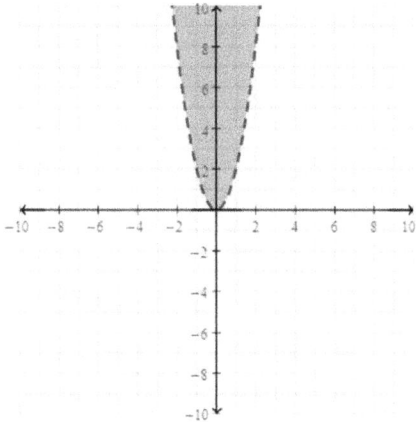

2)

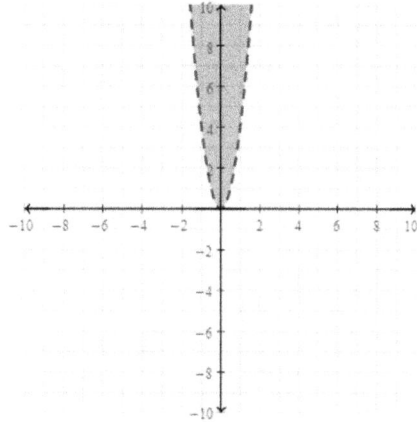

3)

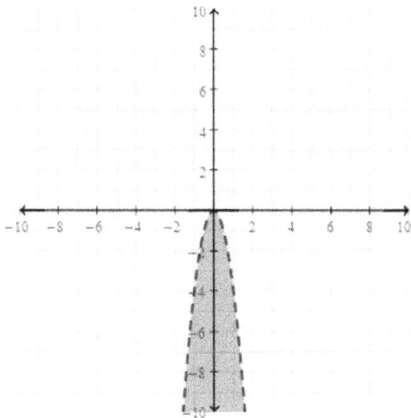

4)

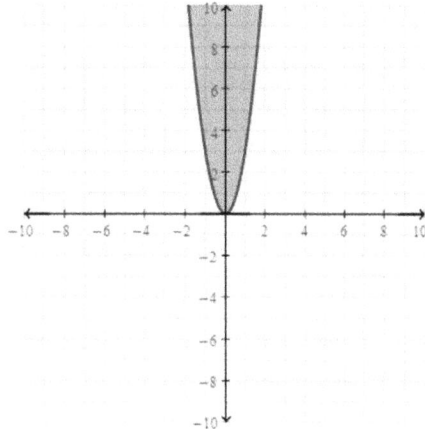

Graphing quadratic functions in vertex form

1)

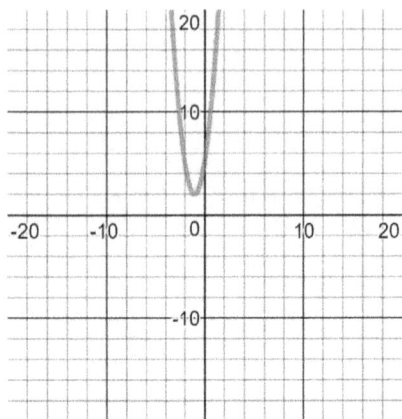

2)

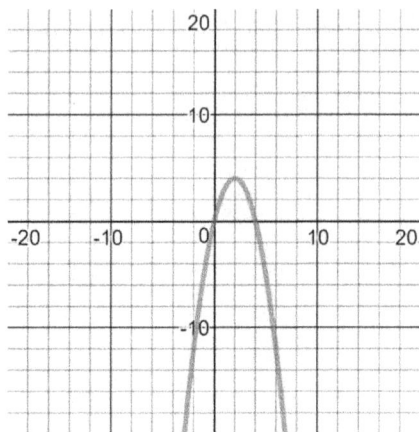

3)

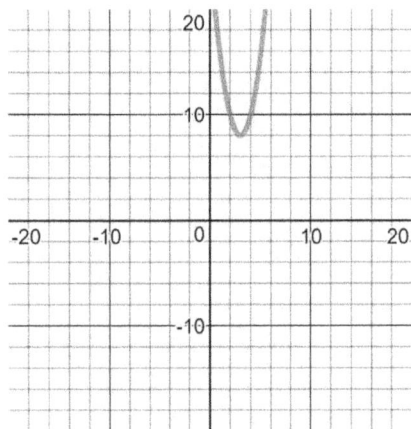

4)

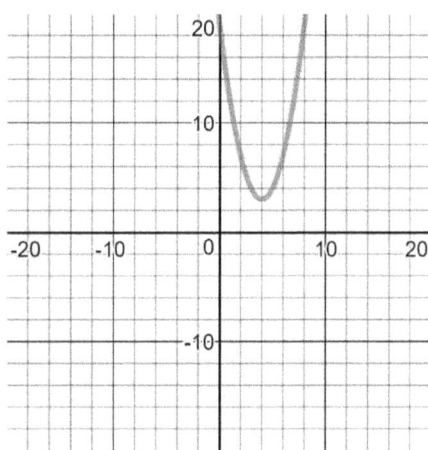

Solving quadratic equations by factoring

1) $\{5, -4\}$

2) $\{-5, -3\}$

3) $\{1\}$

4) $\{4, -1\}$

5) $\{3, -6\}$

6) $\{3, 8\}$

7) $\{-\frac{5}{2}, -\frac{3}{4}\}$

8) $\{-2, 7\}$

9) $\{-3, -5\}$

10) $\{-\frac{7}{5}, -4\}$

11) $\{8, 0\}$

12) $\{\frac{6}{5}, \frac{3}{2}\}$

13) $\{\frac{6}{7}, 0\}$

14) $\{2, 0\}$

15) $\{-\frac{7}{2}, 2\}$

16) $\{\frac{3}{5}, 2\}$

17) $\{-\frac{4}{3}, -4\}$

18) $\{-8, -7\}$

Use the quadratic formula and the discriminant

1) 57

2) 17

3) 41

4) 96

5) −7

6) 76

7) 21

8) 81

9) 13

10) −44

11) 0, one real solution

12) −80, two imaginary solutions

13) 0, one real solution

14) −224, two imaginary solutions

15) 0, one real solution

16) 0, one real solution

17) −63, two imaginary solutions

18) 0, one real solution

Operations with complex numbers

1) 18	7) 62 + 24i	13) – 2 + 4i
2) 18i	8) –56 – 8i	14) 35 – 15i
3) 18 – 4i	9) –125i	15) – 88 + 328i
4) 80 + 48i	10) – 8 + 6i	16) –152 – 384i
5) –7 – 6i	11) – 40 + 40i	17) –48 – 14i
6) –52 + 76i	12) 55 – 48i	18) 4 + 32i

Solve quadratic inequalities

1) $-6 < x < 1$

2) $-1 < x < 6$

3) $x < -5$ or $x > 1$

4) $x \leq -1$ or $x \geq 3$

5) $-1 < x < 1$

6) $x \leq -1$ or $x \geq \frac{2}{17}$

7) $-\frac{11}{2} < x < \frac{1}{2}$

8) $x < -\frac{3}{2}$ or $x > \frac{2}{3}$

9) $-1 \leq x \leq -\frac{5}{18}$

10) $\frac{2}{9} \leq x \leq 3$

11) $x \leq \frac{1}{4}$ or $x \geq \frac{1}{2}$

12) $1 < x < 2$

13) $-3 \leq x \leq -2$

14) $x < -1$ or $x > 6$

15) $x \leq -1$ or $x \geq 3$

16) $x < -3$ or $x > 4$

17) $-3 \leq x \leq -2$

18) $x < -4$ or $x > 2$

Chapter 12: complex numbers

Math Topics that you'll learn today:

- ✓ Adding and Subtracting Complex Numbers

- ✓ Multiplying and Dividing Complex Numbers

- ✓ Graphing Complex Numbers

- ✓ Rationalizing Imaginary Denominators

Mathematics is a hard thing to love. It has the unfortunate habit, like a rude dog, of turning its most unfavorable side towards you when you first make contact with it. — David Whiteland

Adding and Subtracting Complex Numbers

✍ *Simplify.*

1) $- 5 + (2 - 4i)$

2) $- 8 + (2i) + (- 8 + 6i)$

3) $12 - (5i) + (4 - 14i)$

4) $- 2 + (- 8 - 7i) - 9$

5) $(- 18 - 3i) + (11 + 5i)$

6) $(2 - 5i) + (4 - 6i)$

7) $(3 + 5i) + (8 + 3i)$

8) $(8 - 3i) + (4 + i)$

9) $3 + (2 - 4i)$

10) $(10 + 9i) + (6 + 8i)$

11) $(- 5i) - (- 5 + 2i)$

12) $(- 14 + i) - (- 12 - 11i)$

13) $(-12i) + (2 - 6i) + 10$

14) $(-11 - 9i) - (-9 - 3i)$

15) $(13i) - (17 + 3i)$

16) $(- 3 + 6i) - (-9 - i)$

17) $(- 5 + 15i) - (-3 + 3i)$

18) $(- 12i) + (2 - 6i) + 10$

Multiplying and Dividing Complex Numbers

✎ *Simplify.*

1) $(4i)(-i)(2-5i)$

2) $(2-8i)(3-5i)$

3) $(-5+9i)(3+5i)$

4) $(7+3i)(7+8i)$

5) $(5+4i)2$

6) $2(3i)-(5i)(-8+5i)$

7) $\dfrac{2+4i}{14+4i}$

8) $\dfrac{4-3i}{-4i}$

9) $\dfrac{5+6i}{-1+8i}$

10) $\dfrac{-8-i}{-4-6i}$

11) $\dfrac{5+9i}{i}$

12) $\dfrac{12i}{-9+3i}$

13) $\dfrac{5}{-10i}$

14) $\dfrac{-3-10}{5i}$

15) $\dfrac{9i}{3-i}$

16) $\dfrac{-1+5i}{-8-7i}$

17) $\dfrac{-2-9i}{-2+7i}$

18) $\dfrac{4+i}{2-5i}$

Graphing Complex Numbers

✍*Identify each complex number graphed.*

1)

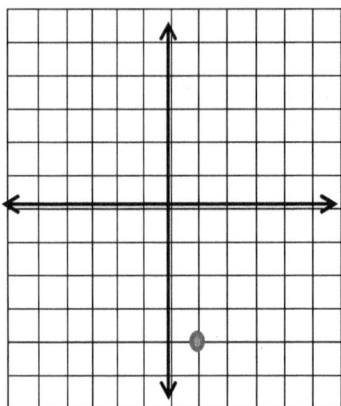

2)

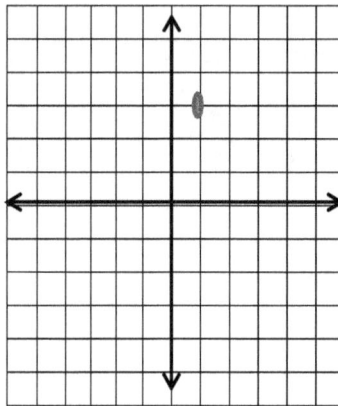

3)

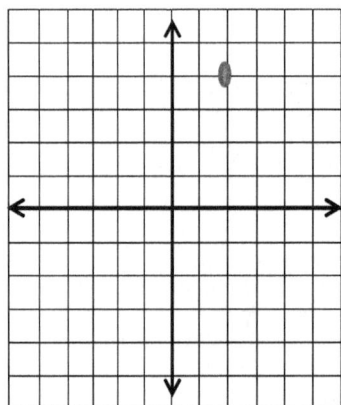

4)

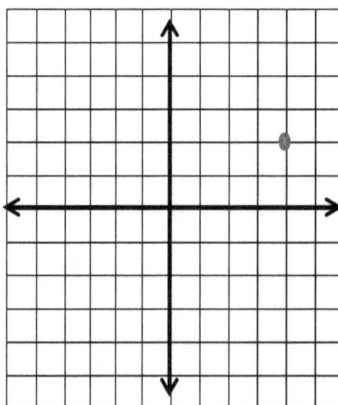

Rationalizing Imaginary Denominators

✍ *Simplify.*

1) $\dfrac{10 - 10i}{-5i}$

2) $\dfrac{4 - 9i}{-6i}$

3) $\dfrac{6 + 8i}{9i}$

4) $\dfrac{8i}{-1+3i}$

5) $\dfrac{5i}{-2 - 6i}$

6) $\dfrac{-10 - 5i}{-6 + 6i}$

7) $\dfrac{-5 - 9i}{9 + 8i}$

8) $\dfrac{-5-3i}{7-10i}$

9) $\dfrac{-1 + i}{-5i}$

10) $\dfrac{-6 - i}{i}$

11) $\dfrac{a}{ib}$

12) $\dfrac{-4 - i}{9 + 5i}$

13) $\dfrac{-3 + i}{-2i}$

14) $\dfrac{-5}{-i}$

15) $\dfrac{-6 - i}{-1 + 6i}$

16) $\dfrac{-9 - 3i}{-3 + 3i}$

17) $\dfrac{6}{-4i}$

18) $\dfrac{8i}{-1 + 3i}$

Answers of Worksheets – Chapter 12

Adding and subtracting complex numbers

1) $-3 - 4i$

2) $-16 + 8i$

3) $16 - 19i$

4) $-19 - 7i$

5) $-7 + 2i$

6) $6 - 11i$

7) $11 + 8i$

8) $12 - 2i$

9) $5 - 4i$

10) $16 + 17i$

11) $5 - 7i$

12) $-2 + 12i$

13) $12 - 18i$

14) $-2 - 6i$

15) $-17 + 10i$

16) $6 + 7i$

17) $-2 + 12i$

18) $12 - 18i$

Multiplying and dividing complex numbers

1) $8 - 20i$

2) $-34 - 34i$

3) $-60 + 2i$

4) $25 + 77i$

5) $9 + 40i$

6) $25 + 46i$

7) $\frac{11+12}{53}$

8) $\frac{3}{4} + i$

9) $\frac{19}{26} - \frac{11}{13}i$

10) $\frac{2-6i}{5}$

11) $-5i + 9$

12) $\frac{-3i+1}{6}$

13) $\frac{i}{2}$

14) $\frac{3i-10}{5}$

15) $\frac{27i-9}{10}$

16) $-\frac{27}{113} - \frac{47}{113}$

17) $-\frac{59}{53} + \frac{32i}{53}$

18) $\frac{3}{29} + \frac{22}{29}$

Graphing complex numbers

1) $1 - 4i$

2) $1 + 3i$

3) $2 + 4i$

4) $4 + 2i$

Rationalizing imaginary denominators

1) $2i + 2$

2) $\dfrac{4 + 9i}{6}$

3) $\dfrac{6+8i}{9}$

4) $\dfrac{-4i+12}{5}$

5) $\dfrac{-i-3}{4}$

6) $\dfrac{5+15i}{12}$

7) $\dfrac{-117-41i}{145}$

8) $\dfrac{-5-71i}{149}$

9) $\dfrac{-i-1}{5}$

10) $6i - 1$

11) $-\dfrac{ia}{b}$

12) $\dfrac{-41+1}{106}$

13) $\dfrac{-1-3i}{2}$

14) $-5i$

15) $0 + 1i$

16) $-3 + 3i$

17) $\dfrac{3i}{2}$

18) $\dfrac{-4i+12}{5}$

Chapter 13: Exponents and Radicals

Math Topics that you'll learn today:

- ✓ Multiplication Property of Exponents
- ✓ Division Property of Exponents
- ✓ Powers of Products and Quotients
- ✓ Zero and Negative Exponents
- ✓ Negative Exponents and Negative Bases
- ✓ Writing Scientific Notation
- ✓ Square Roots

Mathematics is no more computation than typing is literature.

– John Allen Paulos

Multiplication Property of Exponents

✎ *Simplify.*

1) $4^2 \cdot 4^2$

2) $2 \cdot 2^2 \cdot 2^2$

3) $3^2 \cdot 3^2$

4) $3x^3 \cdot x$

5) $12x^4 \cdot 3x$

6) $6x \cdot 2x^2$

7) $5x^4 \cdot 5x^4$

8) $6x^2 \cdot 6x^3y^4$

9) $7x^2y^5 \cdot 9xy^3$

10) $7xy^4 \cdot 4x^3y^3$

11) $(2x^2)^2$

12) $3x^5y^3 \cdot 8x^2y^3$

13) $7x^3 \cdot 10y^3x^5 \cdot 8yx^3$

14) $(x^4)^3$

15) $(2x^2)^4$

16) $(x^2)^3$

17) $(6x)^2$

18) $3x^4y^5 \cdot 7x^2y^3$

Division Property of Exponents

✏️*Simplify.*

1) $\dfrac{5^5}{5}$

2) $\dfrac{3}{3^5}$

3) $\dfrac{2^2}{2^3}$

4) $\dfrac{2^4}{2^2}$

5) $\dfrac{x}{x^3}$

6) $\dfrac{3x^3}{9x^4}$

7) $\dfrac{2x^{-5}}{9x^{-2}}$

8) $\dfrac{21^{\;8}}{7x^3}$

9) $\dfrac{7x^6}{4x^7}$

10) $\dfrac{6x^2}{4x^3}$

11) $\dfrac{5x}{10x^3}$

12) $\dfrac{3x^3}{2x^5}$

13) $\dfrac{12x^3}{14^{\;6}}$

14) $\dfrac{12^{\;3}}{9y^8}$

15) $\dfrac{25xy^4}{5x^6y^2}$

16) $\dfrac{2x^4}{7x}$

17) $\dfrac{16^{\;2}y^8}{4x^3}$

18) $\dfrac{12x^4}{15^{\;7}y^9}$

19) $\dfrac{12yx^4}{10yx^8}$

20) $\dfrac{16x^4y}{9x^8y^2}$

21) $\dfrac{5x^8}{20x^8}$

Powers of Products and Quotients

✏ *Simplify.*

1) $(2x^3)^4$

2) $(4xy^4)^2$

3) $(5x^4)^2$

4) $(11x^5)^2$

5) $(4x^2y^4)^4$

6) $(2x^4y^4)^3$

7) $(3x^2y^2)^2$

8) $(3x^4y^3)^4$

9) $(2x^6y^8)^2$

10) $(12x\ 3x)^3$

11) $(2x^9\ x^6)^3$

12) $(5x^{10}y^3)^3$

13) $(4x^3\ x^2)^2$

14) $(3x^3\ 5x)^2$

15) $(10x^{11}y^3)^2$

16) $(9x^7\ y^5)^2$

17) $(4x^4y^6)^5$

18) $(4x^4)^2$

19) $(3x\ 4y^3)^2$

20) $(9x^2y)^3$

21) $(12x^2y^5)^2$

Zero and Negative Exponents

✎ *Evaluate the following expressions.*

1) 8^{-2}

2) 2^{-4}

3) 10^{-2}

4) 5^{-3}

5) 22^{-1}

6) 9^{-1}

7) 3^{-2}

8) 4^{-2}

9) 5^{-2}

10) 35^{-1}

11) 6^{-3}

12) 0^{15}

13) 10^{-9}

14) 3^{-4}

15) 5^{-2}

16) 2^{-3}

17) 3^{-3}

18) 8^{-1}

19) 7^{-3}

20) 6^{-2}

21) $(\frac{2}{3})^{-2}$

22) $(\frac{1}{5})^{-3}$

23) $(\frac{1}{2})^{-8}$

24) $(\frac{2}{5})^{-3}$

Negative Exponents and Negative Bases

✎ *Simplify.*

1) $- 6^{-1}$

2) $- 4x^{-3}$

3) $- \dfrac{5x}{x^{-3}}$

4) $- \dfrac{a^{-3}}{b^{-2}}$

5) $- \dfrac{5}{x^{-3}}$

6) $\dfrac{7b}{-9c^{-4}}$

7) $- \dfrac{5n^{-2}}{10^{-3}}$

8) $\dfrac{4a^{-2}}{-3c^{-2}}$

9) $- 12x^2y^{-3}$

10) $\left(-\dfrac{1}{3}\right)^{-2}$

11) $\left(-\dfrac{3}{4}\right)^{-2}$

12) $\left(\dfrac{3a}{2c}\right)^{-2}$

13) $\left(-\dfrac{5x}{3y}\right)^{-3}$

14) $- \dfrac{2x}{a^{-4}}$

Writing Scientific Notation

✎ *Write each number in scientific notation.*

1) 91×10^3

2) 60

3) 2000000

4) 0.0000006

5) 354000

6) 0.000325

7) 2.5

8) 0.00023

9) 56000000

10) 2000000

11) 78000000

12) 0.0000022

13) 0.00012

14) 0.004

15) 78

16) 1600

17) 1450

18) 130000

19) 60

20) 0.113

21) 0.02

Square Roots

✎*Find the value each square root.*

1) $\sqrt{1}$

2) $\sqrt{4}$

3) $\sqrt{9}$

4) $\sqrt{25}$

5) $\sqrt{16}$

6) $\sqrt{49}$

7) $\sqrt{36}$

8) $\sqrt{0}$

9) $\sqrt{64}$

10) $\sqrt{81}$

11) $\sqrt{121}$

12) $\sqrt{225}$

13) $\sqrt{144}$

14) $\sqrt{100}$

15) $\sqrt{256}$

16) $\sqrt{289}$

17) $\sqrt{324}$

18) $\sqrt{400}$

19) $\sqrt{900}$

20) $\sqrt{529}$

21) $\sqrt{90}$

Answers of Worksheets – Chapter 13

Multiplication Property of Exponents

1) 4^4
2) 2^5
3) 3^4
4) $3x^4$
5) $36x^5$
6) $12x^3$

7) $25x^8$
8) $36x^5y^4$
9) $63x^3y^8$
10) $28x^4y^7$
11) $4x^4$
12) $24x^7y^6$

13) $560x^{11}y^4$
14) x^{12}
15) $16x^8$
16) x^6
17) $36x^2$
18) $21x^6y^8$

Division Property of Exponents

1) 5^4
2) $\frac{1}{3^4}$
3) $\frac{1}{2}$
4) 2^2
5) $\frac{1}{x^2}$
6) $\frac{1}{3x}$
7) $\frac{2}{9x^3}$
8) $3x^5$

9) $\frac{7}{4x}$
10) $\frac{3}{2x}$
11) $\frac{1}{2x^2}$
12) $\frac{3}{2x^2}$
13) $\frac{6}{7x^3}$
14) $\frac{4x^3}{3y^8}$
15) $\frac{5y^2}{x^5}$

16) $\frac{2x^3}{7}$
17) $\frac{4y^8}{x}$
18) $\frac{4}{5x^3y^9}$
19) $\frac{6}{5x^4}$
20) $\frac{16}{9x^4y}$
21) $\frac{1}{4}$

Powers of Products and Quotients

1) $16x^{12}$
2) $16x^2y^8$
3) $25x^8$
4) $121x^{10}$
5) $256x^8y^{16}$
6) $8x^{12}y^{12}$

7) $9x^4y^4$
8) $81x^{16}y^{12}$
9) $4x^{12}y^{16}$
10) $46,656x^6$
11) $8x^{45}$
12) $125x^{30}y^9$

13) $16x^{10}$
14) $225x^8$
15) $100x^{22}y^6$
16) $81x^{14}y^{10}$
17) $1,024x^{20}y^{30}$
18) $16x^8$

19) $144x^2y^6$ 20) $729x^6y^3$ 21) $144x^4y^{10}$

Zero and Negative Exponents

1) $\frac{1}{64}$

2) $\frac{1}{16}$

3) $\frac{1}{100}$

4) $\frac{1}{125}$

5) $\frac{1}{22}$

6) $\frac{1}{9}$

7) $\frac{1}{9}$

8) $\frac{1}{16}$

9) $\frac{1}{25}$

10) $\frac{1}{35}$

11) $\frac{1}{216}$

12) 0

13) $\frac{1}{1000000000}$

14) $\frac{1}{81}$

15) $\frac{1}{25}$

16) $\frac{1}{8}$

17) $\frac{1}{27}$

18) $\frac{1}{8}$

19) $\frac{1}{343}$

20) $\frac{1}{36}$

21) $\frac{9}{4}$

22) 125

23) 256

24) $\frac{125}{8}$

Negative Exponents and Negative Bases

1) $-\frac{1}{6}$

2) $-\frac{4}{x^3}$

3) $-5x^4$

4) $-\frac{b^2}{a^3}$

5) $-5x^3$

6) $-\frac{7bc^4}{9}$

7) $-\frac{p^3}{2n^2}$

8) $-\frac{4ac^2}{3b^2}$

9) $-\frac{12\ ^2}{y^3}$

10) 9

11) $\frac{16}{9}$

12) $\frac{4c^2}{9a^2}$

13) $-\frac{27\ ^3z^3}{125x^3}$

14) $-2xa^4$

Writing Scientific Notation

1) 9.1×10^4

2) 6×10^1

3) 2×10^6

4) 6×10^{-7}

5) 3.54×10^5

6) 3.25×10^{-4}

7) 2.5×10^0

8) 2.3×10^{-4}

9) 5.6×10^7

10) 2×10^6

11) 7.8×10^7

12) 2.2×10^{-6}

13) 1.2×10^{-4}

14) 4×10^{-3}

15) 7.8×10^1

16) 1.6×10^3

17) 1.45×10^3

18) 1.3×10^5

19) 6×10^1

20) 1.13×10^{-1}

21) 2×10^{-2}

Square Roots

1) 1

2) 2

3) 3

4) 5

5) 4

6) 7

7) 6

8) 0

9) 8

10) 9

11) 11

12) 15

13) 12

14) 10

15) 16

16) 17

17) 18

18) 20

19) 30

20) 23

21) $3\sqrt{10}$

Chapter 14: Statistics

Math Topics that you'll learn today:

- ✓ Mean, Median, Mode, and Range of the Given Data
- ✓ Box and Whisker Plots
- ✓ Bar Graph
- ✓ Stem– And– Leaf Plot
- ✓ The Pie Graph or Circle Graph
- ✓ Scatter Plots

Mathematics is no more computation than typing is literature.

– John Allen Paulos

Mean, Median, Mode, and Range of the Given Data

✍ *Find Mean, Median, Mode, and Range of the Given Data.*

1) 7, 2, 5, 1, 1, 2

2) 2, 2, 2, 3, 6, 3, 7, 4

3) 9, 4, 3, 1, 7, 9, 4, 6, 4

4) 8, 4, 2, 4, 3, 2, 4, 5

5) 8, 5, 7, 5, 7, 9, 8

6) 5, 1, 4, 4, 9, 2, 9, 2, 5, 1

7) 4, 1, 5, 9, 7, 7, 5, 4, 3, 5

8) 7, 5, 4, 9, 6, 7, 7, 5, 2

9) 2, 5, 5, 6, 2, 4, 7, 6, 4, 9

10) 10, 5, 2, 5, 4, 5, 8, 10

11) 5, 1, 5, 2, 2

12) 2, 3, 5, 9, 6

Box and Whisker Plots

Example:

73, 84, 86, 95, 68, 67, 100, 94, 77, 80, 62, 79

Maximum: 100, Minimum: 62, Q_1: 70.5, Q_2: 79.5, Q_3: 90

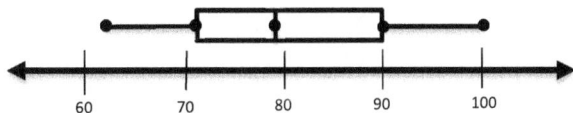

✎*Make box and whisker plots for the given data.*

11, 17, 22, 18, 23, 2, 3, 16, 21, 7, 8, 15, 5

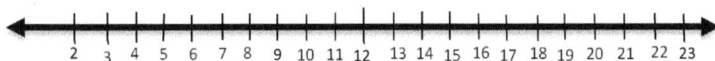

Bar Graph

✎*Graph the given information as a bar graph.*

Day	Hot dogs sold
Monday	90
Tuesday	70
Wednesday	30
Thursday	20
Friday	60

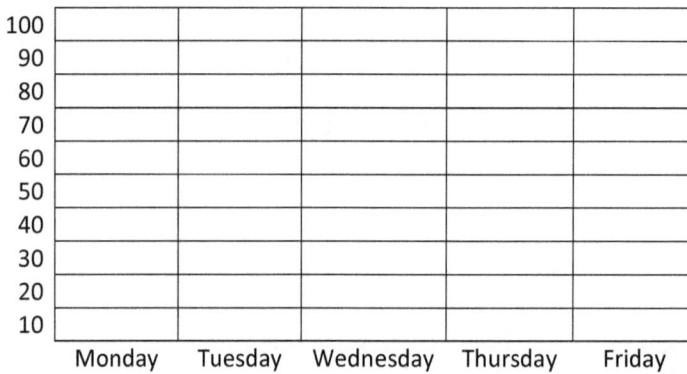

	Monday	Tuesday	Wednesday	Thursday	Friday
100					
90					
80					
70					
60					
50					
40					
30					
20					
10					

Stem–And–Leaf Plot

Example:

56, 58, 42, 48, 66, 64, 53, 69, 45, 72

Stem	leaf		
4	2	5	8
5	3	6	8
6	4	6	9
7	2		

✍ *Make stem ad leaf plots for the given data.*

1) 74, 88, 97, 72, 79, 86, 95, 79, 83, 91

Stem | Leaf plot

2) 37, 48, 26, 33, 49, 26, 19, 26, 48

Stem | Leaf plot

3) 58, 41, 42, 67, 54, 65, 65, 54, 69, 53

Stem | Leaf plot

The Pie Graph or Circle Graph

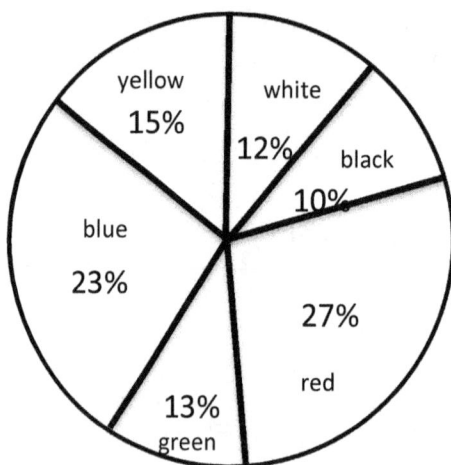

Favorite colors

1) Which color is the most?

2) What percentage of pie graph is yellow?

3) Which color is the least?

4) What percentage of pie graph is blue?

5) What percentage of pie graph is green?

Scatter Plots

✎ *Construct a scatter plot.*

X	Y
1	20
2	40
3	50
4	60

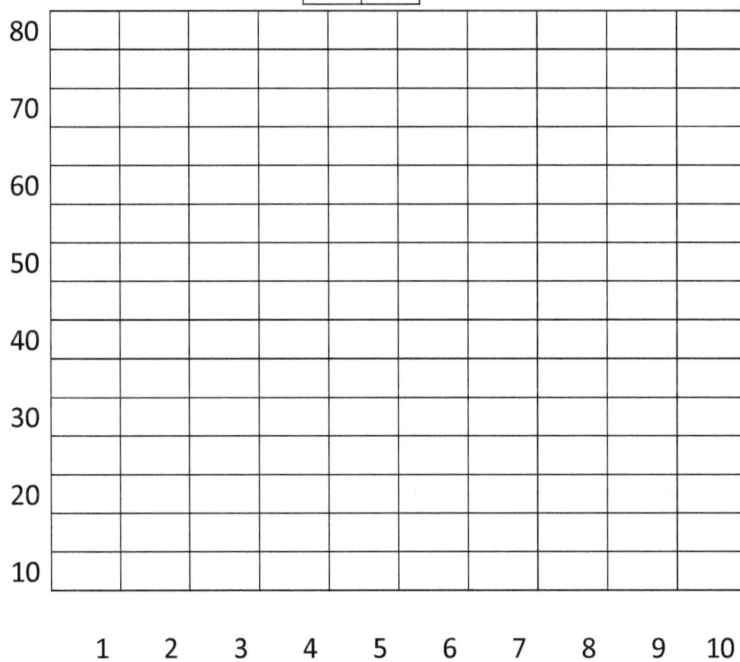

Answers of Worksheets – Chapter 14

Mean, Median, Mode, and Range of the Given Data

1) mean: 3, median: 2, mode: 1, 2, range: 6
2) mean: 3.625, median: 3, mode: 2, range: 5
3) mean: 5.22, median: 4, mode: 4, range: 8
4) mean: 4, median: 4, mode: 4, range: 6
5) mean: 7, median: 7, mode: 5, 7, 8, range: 4
6) mean: 4.2, median: 4, mode: 1,2,4,5,9, range: 8
7) mean: 5, median: 5, mode: 5, range: 8
8) mean: 5.78, median: 6, mode: 7, range: 7
9) mean: 5, median: 5, mode: 2, 4, 5, 6, range: 7
10) mean: 6.125, median: 5, mode: 5, range: 8
11) mean: 3, median: 2, mode: 2, 5, range: 4
12) mean: 5, median: 5, mode: none, range: 7

Box and Whisker Plots

11, 17, 22, 18, 23, 2, 3, 16, 21, 7, 8, 15, 5

Maximum: 23, Minimum: 2, Q_1: 2, Q_2: 12.5, Q_3: 19.5

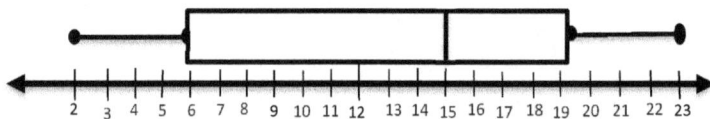

Bar Graph

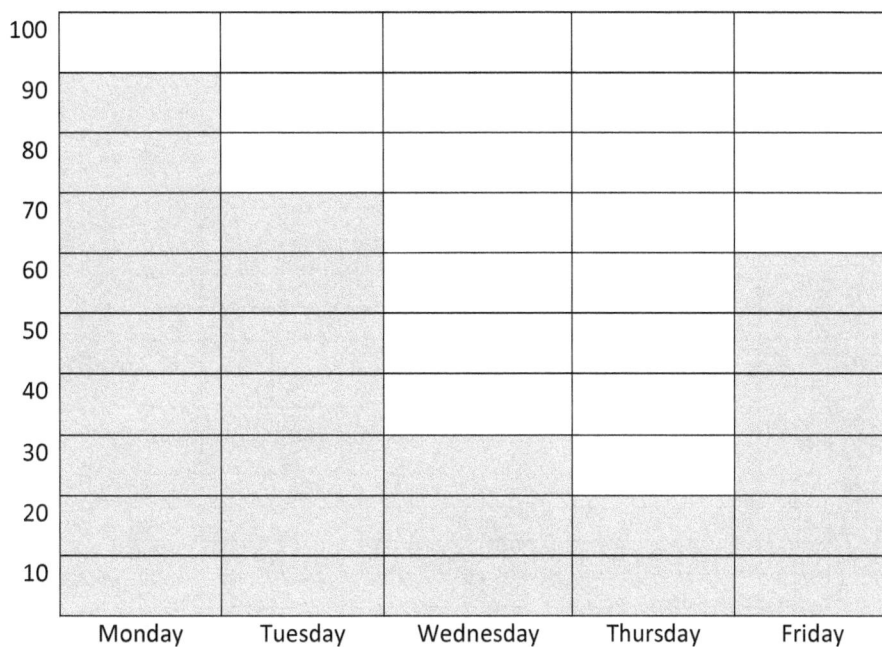

| | Monday | Tuesday | Wednesday | Thursday | Friday |

Stem–And–Leaf Plot

1)

Stem	leaf
7	2 4 9 9
8	3 6 8
9	1 5 7

2)

Stem	leaf
1	9
2	6 6 6
3	3 7
4	8 8 9

3)

Stem	leaf
4	1 2
5	3 4 4 8
6	5 5 7 9

The Pie Graph or Circle Graph

1) red
2) 15%
3) black
4) 23%
5) 13%

Scatter Plots

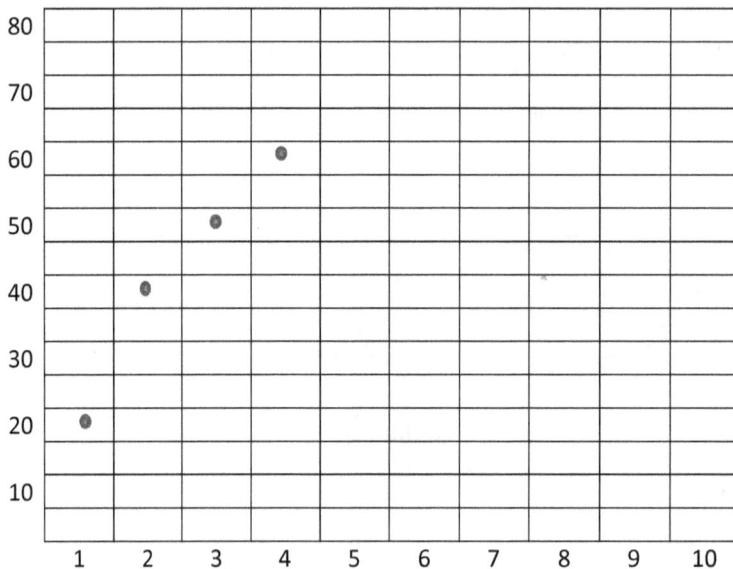

Chapter 15: Geometry

Math Topics that you'll learn today:

- ✓ The Pythagorean Theorem
- ✓ Area of Triangles
- ✓ Perimeter of Polygons
- ✓ Area and Circumference of Circles
- ✓ Area of Squares, Rectangles, and Parallelograms
- ✓ Area of Trapezoids

Mathematics is, as it were, a sensuous logic, and relates to philosophy as do the arts, music, and plastic

art to poetry. — K. Shegel

The Pythagorean Theorem

✍️ *Do the following lengths form a right triangle?*

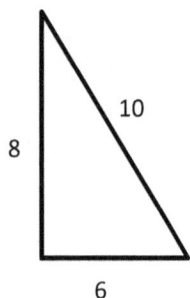

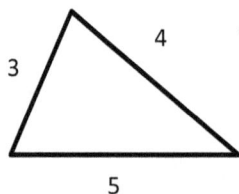

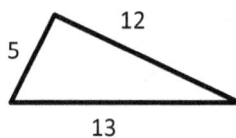

✍️ *Find each missing length to the nearest tenth.*

4)

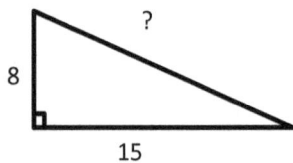

5)

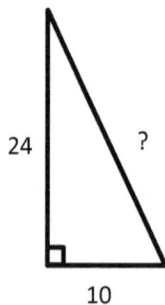

6)

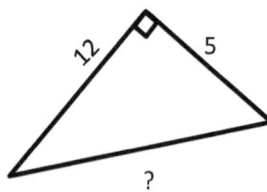

Area of Triangles

🖎*Find the area of each.*

1)

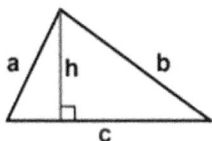

c = 9 mi

h = 3.7 mi

2)

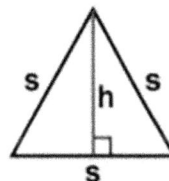

s = 14 m

h = 12.2 m

3)

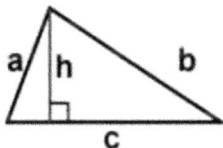

a = 5 m

b = 11 m

c = 14 m

h = 4 m

4)

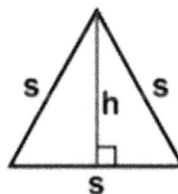

s = 10 m

h = 8.6 m

Perimeter of Polygons

✎ *Find the perimeter of each shape.*

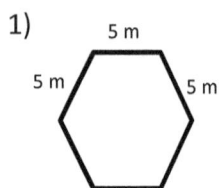

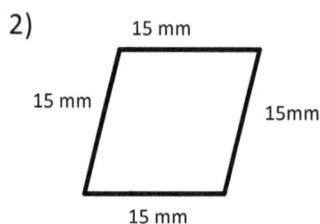

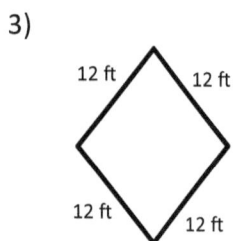

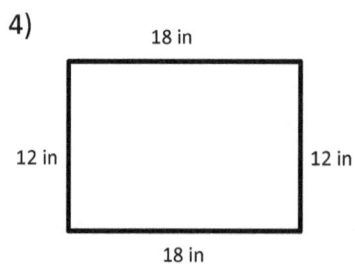

1)

5 m
5 m   5 m

2)

15 mm
15 mm   15mm
15 mm

3)

12 ft   12 ft
12 ft   12 ft

4)

18 in
12 in   12 in
18 in

Area and Circumference of Circles

✎ *Find the area and circumference of each.* ($\pi = 3.14$)

1)

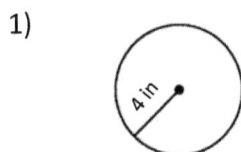

4 in

2)

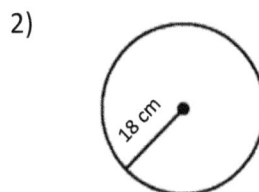

18 cm

3)

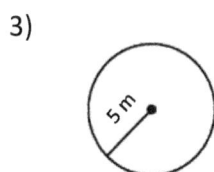

5 m

4)

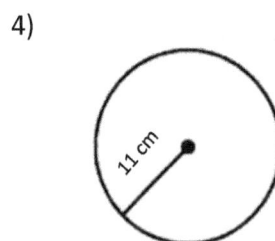

11 cm

5)

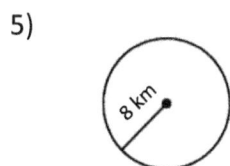

8 km

6)

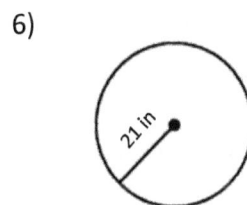

21 in

Area of Squares, Rectangles, and Parallelograms

✍️*Find the area of each.*

1)

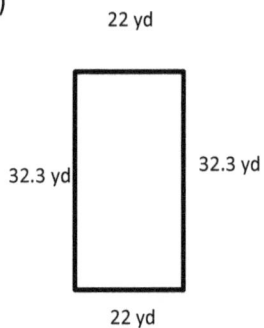

2)

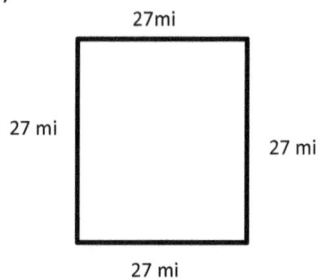

3)

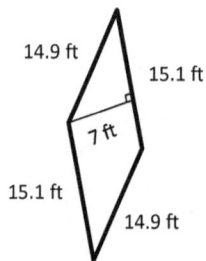

4)

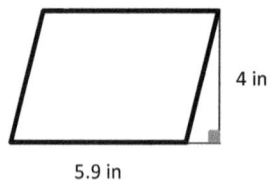

Area of Trapezoids

✏️*Calculate the area for each trapezoid.*

1)

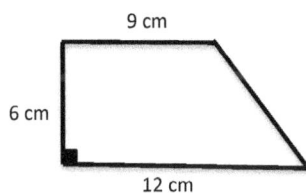

9 cm

6 cm

12 cm

2)

14 m

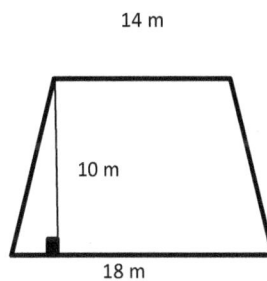

10 m

18 m

3)

22 mi

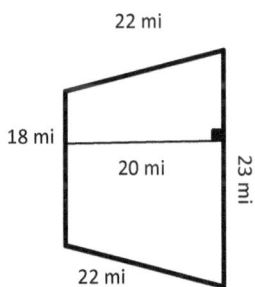

18 mi

20 mi

23 mi

22 mi

4)

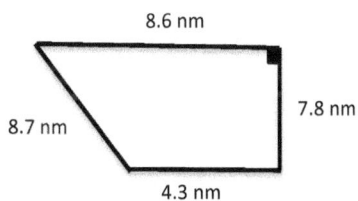

8.6 nm

8.7 nm

7.8 nm

4.3 nm

Answers of Worksheets – Chapter 15

The Pythagorean Theorem

1) yes

2) yes

3) yes

4) 17

5) 26

6) 13

Area of Triangles

1) 16.65 mi^2

2) 56 m^2

3) 85.4 m^2

4) 43 m^2

Perimeter of Polygons

1) 30 m

2) 60 mm

3) 48 ft

4) 60 in

Area and Circumference of Circles

1) Area: 50.24 in^2, Circumference: 25.12 in

2) Area: 1,017.36 cm^2, Circumference: 113.04 cm

3) Area: 78.5m^2, Circumference: 31.4 m

4) Area: 379.94 cm^2, Circumference: 69.08 cm

5) Area: 200.96 km^2, Circumference: 50.2 km

6) Area: 1,384.74 km^2, Circumference: 131.88 km

Area of Squares, Rectangles, and Parallelograms

1) 710.6 yd^2

2) 729 mi^2

3) 105.7 ft^2

4) 23.6 in^2

Area of Trapezoids

1) 63 cm^2

2) 192 m^2

3) 451 mi^2

4) 50.31 nm^2

Chapter 16: Solid Figures

Math Topics that you'll learn today:

- ✓ Volume of Cubes
- ✓ Volume of Rectangle Prisms
- ✓ Surface Area of Cubes
- ✓ Surface Area of Rectangle Prisms
- ✓ Volume of a Cylinder
- ✓ Surface Area of a Cylinder

Mathematics is a ACTat motivator for all humans. Because its career starts with zero and it never

end (infinity)

Volume of Cubes

✍ *Find the volume of each.*

1)

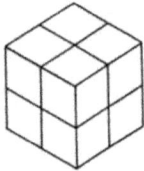

2)

3)

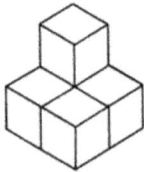

4)

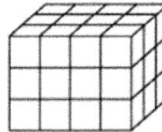

5)

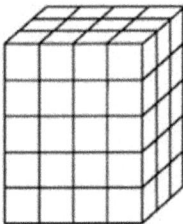

6)

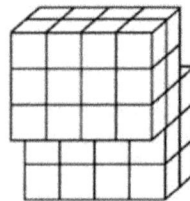

Volume of Rectangle Prisms

✍ *Find the volume of each of the rectangular prisms.*

1)

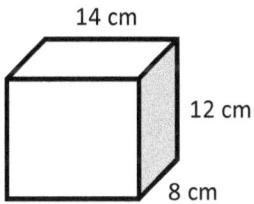

2)

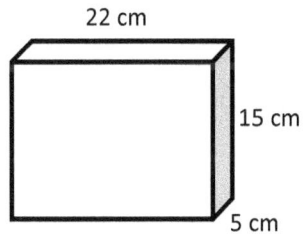

3)

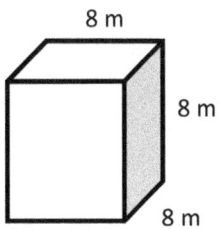

4)

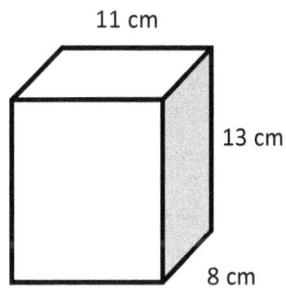

Surface Area of Cubes

✍ *Find the surface of each cube.*

1)

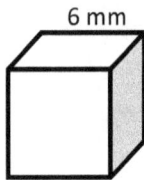

6 mm

2)

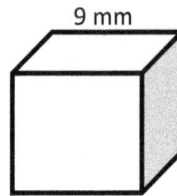

9 mm

3)

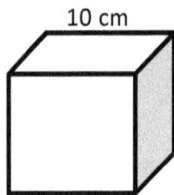

10 cm

4)

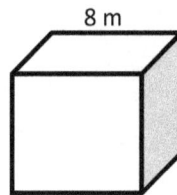

8 m

5)

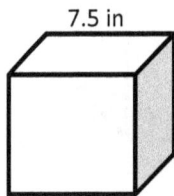

7.5 in

6)

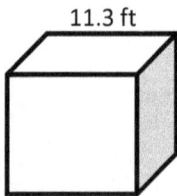

11.3 ft

Surface Area of a Rectangle Prism

✍️*Find the surface of each prism.*

1)

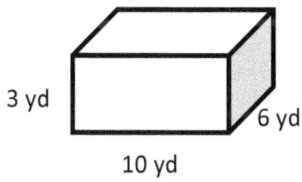

3 yd
6 yd
10 yd

2)

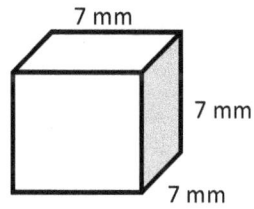

7 mm
7 mm
7 mm

3)

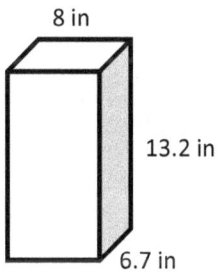

8 in
13.2 in
6.7 in

4)

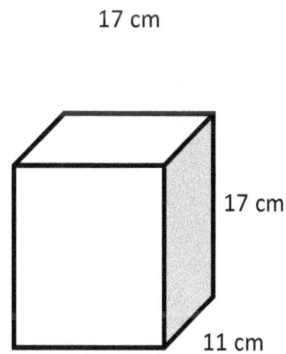

17 cm
17 cm
11 cm

Volume of a Cylinder

✎ *Find the volume of each cylinder.* $(\pi = 3.14)$

1)

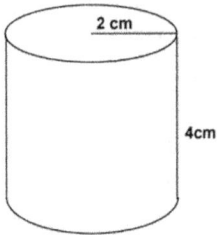

2)

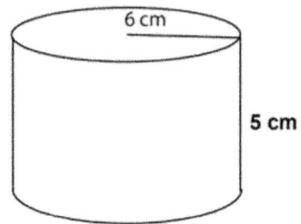

3)

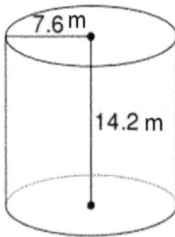

4)

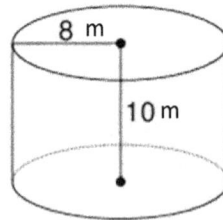

Surface Area of a Cylinder

✎ *Find the surface of each cylinder.* ($\pi = 3.14$)

1)

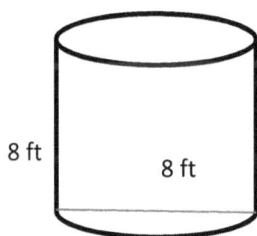

8 ft

8 ft

2)

11 m

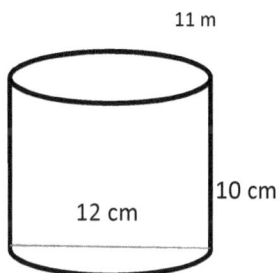

10 cm

12 cm

3)

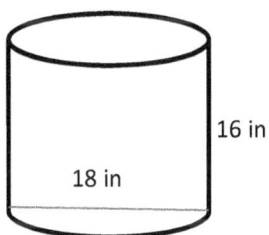

16 in

18 in

4)

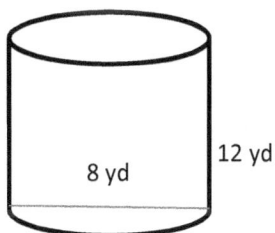

12 yd

8 yd

Answers of Worksheets – Chapter 16

Volumes of Cubes

1) 8

2) 4

3) 5

4) 36

5) 60

6) 44

Volume of Rectangle Prisms

1) 1344 cm^3

2) 1650 cm^3

3) 512 m^3

4) 1144 cm^3

Surface Area of a Cube

1) 216 mm^2

2) 486 mm^2

3) 600 cm^2

4) 384 m^2

5) 337.5 in^2

6) 766.14 ft^2

Surface Area of a Prism

1) 216 yd^2

2) 294 mm^2

3) 495.28 in^2

4) 1326 cm^2

Volume of a Cylinder

1) 50.24 cm^3

2) 565.2 cm^3

3) 2,575.403 m^3

4) 2009.6 m^3

Surface Area of a Cylinder

1) 301.44 ft^2

2) 602.88 cm^2

3) 1413 in^2

4) 401.92 yd^2

Chapter 17: Logarithms

Math Topics that you'll learn today:

- ✓ Rewriting Logarithms
- ✓ Evaluating Logarithms
- ✓ Properties of Logarithms
- ✓ Natural Logarithms
- ✓ Solving Exponential Equations Requiring Logarithms
- ✓ Solving Logarithmic Equations

Mathematics is an art of human understanding. — William Thurston

Rewriting Logarithms

✍ *Rewrite each equation in exponential form.*

1) $\log_{24} 15 = 0.85$

2) $\log_{320} 35 = 0.61$

3) $\log_7 49 = 2$

4) $\log_6 36 = 2$

✍ *Rewrite each equation in exponential form.*

5) $\log_a \frac{5}{8} = b$

6) $\log_x y = 6$

7) $\log_{12} n = m$

8) $\log_y x = -8$

9) $\log_a b = 22$

10) $\log_{\frac{1}{5}} v = u$

✍ *Evaluate each expression.*

11) $\log_4 64$

12) $\log_4 16$

13) $\log_5 125$

14) $\log_9 3$

Evaluating Logarithms

✎ *Evaluate each expression.*

1) $\log_3 27$

2) $\log_2 32$

3) $\log_4 16$

4) $\log_2 4$

5) $\log_8 64$

6) $\log_7 \frac{1}{49}$

7) $\log_{64} \frac{1}{4}$

8) $\log_{80} 700$

9) $\log_4 \frac{1}{64}$

10) $\log_5 625$

11) $\log_6 216$

12) $\log_8 \frac{1}{216}$

13) $\log_8 512$

14) $\log_7 2401$

Properties of Logarithms

✎ *Expand each logarithm.*

1) $\log \left(\frac{2}{5}\right)^3$

2) $\log (2 . 3^4)$

3) $\log \left(\frac{5}{7}\right)^4$

4) $\log \frac{2^3}{7}$

5) $\log (x . y)^5$

6) $\log (8 . 5)$

7) $\log (3 . 7)$

8) $\log (x^3 . y . z^4)$

9) $\log \frac{u^4}{v}$

10) $\log \frac{x}{y^6}$

✎ *Condense each expression to a single logarithm.*

11) $\log 2 - \log 9$

12) $5 \log 6 - 3 \log 4$

13) $\log 7 - 2 \log 12$

14) $4 \log_5 a + 7 \log_5 b$

15) $2\log_3 x - 9 \log_3 y$

16) $\log_4 u - 6 \log_4 v$

17) $4 \log_6 u + 8 \log_6 v$

18) $4 \log_3 u - 20 \log_3 v$

Natural Logarithms

✎ *Solve.*

1) $e^x = 3$

2) $\ln(\ln x) = 5$

3) $e^x = 9$

4) $\ln(2x + 5) = 4$

5) $\ln(6x - 1) = 1$

6) $\ln x = \dfrac{1}{2}$

7) $x = e^{\frac{1}{2}}$

8) $\ln x = \ln 4 + \ln 7$

✎ *Evaluate without using a calculator.*

9) $\ln 1$

10) $\ln e^3$

11) $4 \ln e$

12) $\ln \left(\dfrac{1}{e}\right)$

13) $e^{\ln 10}$

14) $e^{3\ln 2}$

15) $e^{5\ln 2}$

16) $\ln \sqrt{e}$

Solving Exponential Equations Requiring Logarithms

Solve each equation.

1) $4^{r+1} = 1$

2) $243^x = 81$

3) $6^{-3v-2} = 36$

4) $3^{2n} = 9$

5) $\frac{216^{2a}}{36^{-a}} = 216$

6) $25 \cdot 25^{-v} = 625$

7) $3^{2n} = 9$

8) $(\frac{1}{6})^n = 36$

9) $32^{2x} = 8$

10) $2^{-3x} = 2^{x-1}$

11) $2^{2n} = 16$

12) $5^{3n} = 125$

13) $3^{-2k} = 81$

14) $5^{3r} = 5^{-2r}$

15) $10^{3x} = 10000$

16) $25 \cdot 125^{-v} = 625$

17) $\frac{125}{25^{-3m}} = 25^{-2m-2}$

18) $2^{-2n} \cdot 2^{n+1} = 2^{-2n}$

Solving Logarithmic Equations

✍ *Solve each equation.*

1) $2 \log_7 - 2x = 0$

2) $- \log_5 7x = 2$

3) $\log x + 5 = 2$

4) $\log x - \log 4 = 3$

5) $\log x + \log 2 = 4$

6) $\log 10 + \log x = 1$

7) $\log x + \log 8 = \log 48$

8) $- 3 \log_3 (x - 2) = - 12$

9) $\log 6x = \log (x + 5)$

10) $\log (4k - 5) = \log (2k - 1)$

11) $\log (4p - 2) = \log (-5p + 5)$

12) $-10 + \log_3 (n + 3) = -10$

13) $\log_9 (x + 2) = \log_9 (x^2 + 30)$

14) $\log_{12} (v^2 + 35) = \log_{12} (-2v - 1)$

15) $\log (16 + 2b) = \log (b^2 - 4b)$

16) $\log_9 (x + 6) - \log_9 x = \log_9 2$

17) $\log_5 6 + \log_5 2x^2 = \log_5 48$

18) $\log_6 (x + 1) - \log_6 x = \log_6 29$

Answers of Worksheets – Chapter 17

Rewriting logarithms

1) $24^{0.85} = 15$

2) $320^{0.61} = 35$

3) $7^2 = 49$

4) $6^2 = 36$

5) $a^b = \frac{5}{8}$

6) $x^6 = y$

7) $12^m = n$

8) $y^{-8} = x$

9) $a^{22} = b$

10) $(\frac{1}{5})^u = v$

11) 3

12) 2

13) 3

14) $\frac{1}{2}$

Evaluating logarithms

1) 3

2) 5

3) 2

4) 2

5) 2

6) −2

7) $-\frac{1}{3}$

8) 1.5

9) −3

10) 4

11) 3

12) −3

13) 3

14) 4

Properties of logarithms

1) $3 \log 2 - 3 \log 5$
2) $\log 2 + 4 \log 3$
3) $4\log 5 - 4 \log 7$
4) $3 \log 2 - \log 7$
5) $5 \log x + 5 \log y$
6) $Log\ 8 + \log 5$
7) $Log\ 3 + \log 7$
8) $3Log\ x + \log y + 4 \log z$
9) $4 \log u - \log v$
10) $Log\ x - 6 \log y$

11) $\log \frac{2}{9}$

12) $\log \frac{6^5}{4^3}$

13) $\log \frac{7}{12^2}$

14) $\log_5 (a^4 b^7)$

15) $\log_3 \frac{x^2}{y^9}$

16) $\log_4 \frac{u}{v^6}$

17) $\log_6 (v^8 u^4)$

18) $\log_3 \frac{u^4}{v^{20}}$

Natural logarithms

1) $x = \ln 3$
2) $x = e^{e^5}$
3) $x = \ln 9$
4) $x = \frac{e^2 - 5}{2}$
5) $x = \frac{e + 1}{6}$

6) $\ln(e^{3-x}) = 8$
7) $x = -5$
8) $x = 28$
9) 0
10) 3
11) 4

12) -1
13) 10
14) 8
15) 32
16) $\frac{1}{2}$

Solving exponential equations requiring logarithms

1) $-\frac{1}{2}$
2) $\frac{1}{48}$
3) 95
4) 4000
5) 50
6) 1

7) 6
8) 83
9) 1
10) $\frac{1}{4}$
11) 2
12) 1

13) -2
14) 0
15) $\frac{4}{3}$
16) $-\frac{2}{3}$
17) $-\frac{7}{10}$
18) -1

Solving logarithmic equations

1) $\{-\frac{1}{2}\}$
2) $\{\frac{1}{35}\}$
3) $\{-100\}$
4) $\{20\}$
5) $\{\frac{25}{2}\}$
6) $\{5\}$

7) $\{\frac{37}{7}\}$
8) $\{84\}$
9) $\{3\}$
10) $\{2\}$
11) $\{\frac{7}{9}\}$
12) $\{-2\}$

13) $\{-7, -4\}$
14) $\{-6\}$
15) $\{8, -2\}$
16) $\{6\}$
17) $\{2, -2\}$
18) $\{\frac{1}{28}\}$

Chapter 18: Matrices

Math Topics that you'll learn today:

✓ Adding and Subtracting Matrices

✓ Matrix Multiplications

✓ Finding Determinants of a Matrix

✓ Finding Inverse of a Matrix

✓ Matrix Equations

Mathematics is an independent world created out of pure intelligence.

— William Woods Worth

Adding and Subtracting Matrices

✎ *Simplify.*

1) $|2 \quad -5 \quad -3| + |1 \quad -2 \quad -3|$

2) $\begin{vmatrix} 3 & 6 \\ -1 & -3 \\ -5 & -1 \end{vmatrix} + \begin{vmatrix} 0 & -1 \\ 6 & 0 \\ 2 & 3 \end{vmatrix}$

3) $\begin{vmatrix} -5 & 2 & -2 \\ 4 & -2 & 0 \end{vmatrix} - \begin{vmatrix} 6 & -5 & -6 \\ 1 & 3 & -3 \end{vmatrix}$

4) $|4 \quad 2| + |-2 \quad -6|$

5) $\begin{vmatrix} 2 \\ 4 \end{vmatrix} + \begin{vmatrix} 5 \\ 6 \end{vmatrix}$

6) $\begin{vmatrix} -4n & n+m \\ -2n & -4m \end{vmatrix} + \begin{vmatrix} 4 & -5 \\ 3m & 0 \end{vmatrix}$

7) $\begin{vmatrix} -6r+t \\ -r \\ 6s \end{vmatrix} + \begin{vmatrix} 6r \\ -4t \\ -3r+2 \end{vmatrix}$

8) $\begin{vmatrix} z-5 \\ -6 \\ -1-6z \\ 3y \end{vmatrix} + \begin{vmatrix} -3y \\ 3z \\ 5+z \\ 4z \end{vmatrix}$

9) $\begin{vmatrix} 8 & 7 \\ -6 & 5 \end{vmatrix} + \begin{vmatrix} 4 & -3 \\ 1 & 13 \end{vmatrix}$

10) $|-13 \quad 18 \quad 12| + |34 \quad -3 \quad 9|$

11) $\begin{vmatrix} 2 & -5 & 9 \\ 4 & -7 & 11 \\ -6 & 3 & -17 \end{vmatrix} + \begin{vmatrix} 3 & 4 & -5 \\ 13 & 2 & 5 \\ 4 & -8 & 1 \end{vmatrix}$

12) $\begin{vmatrix} 1 & -7 & 15 \\ 31 & 3 & 18 \\ 22 & 6 & 4 \end{vmatrix} + \begin{vmatrix} 13 & 17 & 5 \\ 3 & 8 & -1 \\ -9 & 2 & 12 \end{vmatrix}$

Matrix Multiplication

✎ *Simplify.*

1) $\begin{vmatrix} -5 & -5 \\ -1 & 2 \end{vmatrix} \cdot \begin{vmatrix} -2 & -3 \\ 3 & 5 \end{vmatrix}$

2) $\begin{vmatrix} 0 & 5 \\ -3 & 1 \\ -5 & 1 \end{vmatrix} \cdot \begin{vmatrix} -4 & 4 \\ -2 & -4 \end{vmatrix}$

3) $\begin{vmatrix} 3 & 2 & 5 \\ 2 & 3 & 1 \end{vmatrix} \cdot \begin{vmatrix} 4 & 5 & -5 \\ 5 & -1 & 6 \end{vmatrix}$

4) $\begin{vmatrix} -5 \\ 6 \\ 0 \end{vmatrix} \cdot \begin{vmatrix} 3 & -1 \end{vmatrix}$

5) $\begin{vmatrix} 3 & -1 \\ -3 & 6 \\ -6 & -6 \end{vmatrix} \cdot \begin{vmatrix} -1 & 6 \\ 5 & 4 \end{vmatrix}$

6) $\begin{vmatrix} -2 & -6 \\ -4 & 3 \\ 5 & 0 \\ 4 & -6 \end{vmatrix} \cdot \begin{vmatrix} 2 & -2 & 2 \\ -2 & 0 & -3 \end{vmatrix}$

7) $\begin{vmatrix} -4 & -y \\ -2x & -4 \end{vmatrix} \cdot \begin{vmatrix} -4x & 0 \\ 2y & -5 \end{vmatrix}$

8) $\begin{vmatrix} 2 & -5v \end{vmatrix} \cdot \begin{vmatrix} -5u & -v \\ 0 & 6 \end{vmatrix}$

9) $\begin{vmatrix} -1 & 1 & -1 \\ 5 & 2 & -5 \\ 6 & -5 & 1 \\ -5 & 6 & 0 \end{vmatrix} \cdot \begin{vmatrix} 6 & 5 \\ 5 & -6 \\ 6 & 0 \end{vmatrix}$

10) $\begin{vmatrix} 5 & 3 & 5 \\ 1 & 5 & 0 \end{vmatrix} \cdot \begin{vmatrix} -4 & 2 \\ -3 & 4 \\ 3 & -5 \end{vmatrix}$

11) $\begin{vmatrix} -3 & 5 \\ -2 & 1 \end{vmatrix} \cdot \begin{vmatrix} 6 & -2 \\ 1 & -5 \end{vmatrix}$

12) $\begin{vmatrix} 0 & 2 \\ -2 & -5 \end{vmatrix} \cdot \begin{vmatrix} 6 & -6 \\ 3 & 0 \end{vmatrix}$

Finding Determinants of a Matrix

✎ *Evaluate the determinant of each matrix.*

1) $\begin{vmatrix} 0 & -4 \\ -6 & -2 \end{vmatrix}$

9) $\begin{vmatrix} 0 & 6 \\ -6 & 0 \end{vmatrix}$

2) $\begin{vmatrix} 5 & 3 \\ 6 & 6 \end{vmatrix}$

10) $\begin{vmatrix} 0 & 4 \\ 6 & 5 \end{vmatrix}$

3) $\begin{vmatrix} -1 & 1 \\ -1 & 4 \end{vmatrix}$

11) $\begin{vmatrix} -2 & 5 & -4 \\ 0 & -3 & 5 \\ -5 & 5 & -6 \end{vmatrix}$

4) $\begin{vmatrix} -9 & -9 \\ -7 & -10 \end{vmatrix}$

12) $\begin{vmatrix} 5 & 3 & 3 \\ -4 & -5 & 1 \\ 5 & 3 & 0 \end{vmatrix}$

5) $\begin{vmatrix} -1 & 8 \\ 5 & 0 \end{vmatrix}$

13) $\begin{vmatrix} 6 & 2 & -1 \\ -5 & -4 & -5 \\ 3 & -3 & 1 \end{vmatrix}$

6) $\begin{vmatrix} 8 & -6 \\ -10 & 9 \end{vmatrix}$

14) $\begin{vmatrix} 6 & 5 & -3 \\ -5 & 4 & -2 \\ 1 & -4 & 5 \end{vmatrix}$

7) $\begin{vmatrix} 2 & -2 \\ 7 & -7 \end{vmatrix}$

8) $\begin{vmatrix} -5 & 0 \\ 3 & 10 \end{vmatrix}$

15) $\begin{vmatrix} -1 & -8 & 9 \\ 4 & 12 & -7 \\ -10 & 3 & 2 \end{vmatrix}$

Finding Inverse of a Matrix

✎ *Find the inverse of each matrix.*

1) $\begin{vmatrix} 3 & -2 \\ -4 & 6 \end{vmatrix}$

2) $\begin{vmatrix} 5 & -8 \\ 6 & -9 \end{vmatrix}$

3) $\begin{vmatrix} 2 & -10 \\ -11 & 8 \end{vmatrix}$

4) $\begin{vmatrix} -9 & -6 \\ -5 & -4 \end{vmatrix}$

5) $\begin{vmatrix} -3 & 3 \\ 8 & 7 \end{vmatrix}$

6) $\begin{vmatrix} -2 & 2 \\ -9 & 8 \end{vmatrix}$

7) $\begin{vmatrix} 3 & -2 \\ -4 & 6 \end{vmatrix}$

8) $\begin{vmatrix} -6 & 11 \\ -4 & 7 \end{vmatrix}$

9) $\begin{vmatrix} -1 & 7 \\ -1 & 7 \end{vmatrix}$

10) $\begin{vmatrix} 1 & -1 \\ -6 & -3 \end{vmatrix}$

11) $\begin{vmatrix} 11 & -5 \\ 2 & -1 \end{vmatrix}$

12) $\begin{vmatrix} 0 & -2 \\ -1 & -9 \end{vmatrix}$

13) $\begin{vmatrix} 0 & 0 \\ -6 & 4 \end{vmatrix}$

14) $\begin{vmatrix} -9 & -9 \\ -2 & -2 \end{vmatrix}$

Matrix Equations

✎ *Solve each equation.*

1) $\begin{vmatrix} -1 & 2 \\ -6 & 10 \end{vmatrix} z = \begin{vmatrix} 6 \\ 22 \end{vmatrix}$

7) $\begin{vmatrix} -1 & 1 \\ 5 & -2 \end{vmatrix} c = \begin{vmatrix} 4 \\ -26 \end{vmatrix}$

2) $3x = \begin{vmatrix} 12 & -12 \\ 21 & -27 \end{vmatrix}$

8) $\begin{vmatrix} 4 & -2 \\ -7 & 2 \end{vmatrix} c = \begin{vmatrix} -6 \\ 12 \end{vmatrix}$

3) $\begin{vmatrix} 20 & -3 \\ 15 & -3 \end{vmatrix} = \begin{vmatrix} -6 & -5 \\ -5 & -4 \end{vmatrix} x$

9) $\begin{vmatrix} 2 & -3 \\ -5 & 5 \end{vmatrix} z = \begin{vmatrix} -1 \\ 20 \end{vmatrix}$

4) $Y - \begin{vmatrix} -1 \\ -5 \\ 8 \\ 8 \end{vmatrix} = \begin{vmatrix} -6 \\ 6 \\ -16 \\ 0 \end{vmatrix}$

10) $\begin{vmatrix} -5 \\ 5 \\ -20 \end{vmatrix} = 5B$

11) $\begin{vmatrix} -10 \\ 4 \\ 3 \end{vmatrix} = y - \begin{vmatrix} 7 \\ -5 \\ -11 \end{vmatrix}$

5) $\begin{vmatrix} -1 & -9 \\ 0 & -1 \end{vmatrix} c = \begin{vmatrix} 11 \\ 2 \end{vmatrix}$

6) $\begin{vmatrix} -1 & -2 \\ 2 & 9 \end{vmatrix} B = \begin{vmatrix} -3 & -5 & 13 \\ 21 & 0 & -36 \end{vmatrix}$

12) $-4b - \begin{vmatrix} 5 \\ 2 \\ -6 \end{vmatrix} = \begin{vmatrix} -33 \\ -2 \\ -22 \end{vmatrix}$

Answers of Worksheets – Chapter 18

Adding and subtracting matrices

1) $\begin{vmatrix} 3 & -7 & -6 \end{vmatrix}$

2) $\begin{vmatrix} 3 & 5 \\ 5 & -3 \\ -3 & 2 \end{vmatrix}$

3) $\begin{vmatrix} -11 & 7 & 4 \\ 3 & -5 & 3 \end{vmatrix}$

4) $\begin{vmatrix} 2 & -4 \end{vmatrix}$

5) $\begin{vmatrix} 7 \\ 10 \end{vmatrix}$

6) $\begin{vmatrix} -4n+4 & n+m-5 \\ -2n+3m & -4m \end{vmatrix}$

7) $\begin{vmatrix} t \\ -r-4t \\ 6s-3r+2 \end{vmatrix}$

8) $\begin{vmatrix} z-5-3y \\ -6+3z \\ -4-5z \\ 3y+4z \end{vmatrix}$

9) $\begin{vmatrix} 12 & 4 \\ -5 & 18 \end{vmatrix}$

10) $\begin{vmatrix} 21 & 15 & 21 \end{vmatrix}$

11) $\begin{vmatrix} -1 & -9 & 14 \\ -9 & -9 & 6 \\ -6 & 11 & -18 \end{vmatrix}$

12) $\begin{vmatrix} 14 & 10 & 20 \\ 34 & 11 & 17 \\ 13 & 8 & 16 \end{vmatrix}$

Matrix multiplication

1) $\begin{vmatrix} -5 & -10 \\ 8 & 13 \end{vmatrix}$

2) $\begin{vmatrix} -10 & -20 \\ 10 & -16 \\ 18 & -24 \end{vmatrix}$

3) Undefined

4) $\begin{vmatrix} -15 & 5 \\ 18 & -6 \\ 0 & 0 \end{vmatrix}$

5) $\begin{vmatrix} -8 & 14 \\ 33 & 6 \\ -24 & -60 \end{vmatrix}$

6) $\begin{vmatrix} 8 & 4 & 14 \\ -14 & 8 & -17 \\ 10 & -10 & 10 \\ 20 & -8 & 26 \end{vmatrix}$

9) $\begin{vmatrix} -7 & -11 \\ 10 & 13 \\ 17 & 60 \\ 0 & -61 \end{vmatrix}$

7) $\begin{vmatrix} 16x - 2y^2 & 5y \\ 8x^2 - 8y & 20 \end{vmatrix}$

10) $\begin{vmatrix} -14 & -3 \\ -19 & 22 \end{vmatrix}$

8) $\begin{vmatrix} -10u & -32v \end{vmatrix}$

11) $\begin{vmatrix} -13 & -19 \\ -11 & -1 \end{vmatrix}$

12) $\begin{vmatrix} 6 & 0 \\ -27 & 12 \end{vmatrix}$

Finding determinants of a matrix

1) −24

2) 12

3) −3

4) 27

5) −40

6) 12

7) 0

8) −50

9) −36

10) −24

11) −51

12) 39

13) −161

14) 139

15) 647

Finding inverse of a matrix

1) $\begin{vmatrix} \frac{3}{5} & \frac{1}{5} \\ \frac{2}{5} & \frac{3}{10} \end{vmatrix}$

2) $\begin{vmatrix} -3 & \frac{8}{3} \\ -2 & \frac{5}{3} \end{vmatrix}$

3) $\begin{vmatrix} -\frac{4}{47} & -\frac{5}{47} \\ -\frac{2}{94} & -\frac{1}{47} \end{vmatrix}$

4) $\begin{vmatrix} -\frac{2}{3} & 1 \\ \frac{5}{6} & -\frac{3}{2} \end{vmatrix}$

5) $\begin{vmatrix} -\frac{7}{45} & \frac{1}{15} \\ \frac{8}{45} & \frac{1}{15} \end{vmatrix}$

6) $\begin{vmatrix} 4 & -1 \\ \frac{9}{2} & -1 \end{vmatrix}$

7) $\begin{vmatrix} \frac{3}{5} & \frac{1}{5} \\ \frac{2}{5} & \frac{3}{10} \end{vmatrix}$

8) $\begin{vmatrix} \frac{7}{2} & -\frac{11}{2} \\ 2 & -3 \end{vmatrix}$

9) No inverse exists

10) $\begin{vmatrix} \frac{1}{3} & -\frac{1}{9} \\ -\frac{2}{3} & -\frac{1}{9} \end{vmatrix}$

11) $\begin{vmatrix} 1 & -5 \\ 2 & -11 \end{vmatrix}$

12) $\begin{vmatrix} \frac{9}{2} & -1 \\ -\frac{1}{2} & 0 \end{vmatrix}$

13) No inverse exists

14) No inverse exists

Matrix equations

1) $\begin{vmatrix} 8 \\ 7 \end{vmatrix}$

2) $\begin{vmatrix} 4 & -4 \\ 7 & -9 \end{vmatrix}$

3) $\begin{vmatrix} 5 & 3 \\ -10 & -3 \end{vmatrix}$

4) $\begin{vmatrix} -7 \\ 1 \\ -8 \\ 8 \end{vmatrix}$

5) $\begin{vmatrix} 7 \\ -2 \end{vmatrix}$

6) $\begin{vmatrix} -3 & 9 & -9 \\ 3 & -2 & -2 \end{vmatrix}$

7) $\begin{vmatrix} -6 \\ -2 \end{vmatrix}$

8) $\begin{vmatrix} -2 \\ -1 \end{vmatrix}$

9) $\begin{vmatrix} -11 \\ -7 \end{vmatrix}$

10) $\begin{vmatrix} -1 \\ 1 \\ -4 \end{vmatrix}$

11) $\begin{vmatrix} -3 \\ -1 \\ -8 \end{vmatrix}$

12) $\begin{vmatrix} 7 \\ 0 \\ 7 \end{vmatrix}$

Chapter 19: Functions Operations

Math Topics that you'll learn today:

- ✓ Relations and Functions
- ✓ Function Notation
- ✓ Adding and Subtracting Functions
- ✓ Multiplying and Dividing Functions
- ✓ Composition of Functions

Millions saw the apple fall, but Newton asked why." – Bernard Baruch

Function Notation

✍ *Write in function notation.*

1) $d = 22t$

2) $c = p^2 + 5p + 5$

3) $m = 25n - 120$

4) $y = 2x - 6$

✍ *Evaluate each function.*

5) $w(x) = 3x + 1$, find $w(4)$

6) $h(n) = n^2 - 10$, find $h(5)$

7) $h(x) = x^3 + 8$, find $h(-2)$

8) $h(n) = -2n^2 - 6n$, find $h(2)$

9) $g(n) = 3n^2 + 2n$, find $g(2)$

10) $g(n) = 10n - 3$, find $g(6)$

11) $g(n) = 8n + 4$, find $g(1)$

12) $h(x) = 4x - 22$, find $h(2)$

13) $h(a) = -11a + 5$, find $h(2a)$

14) $k(a) = 7a + 3$, find $k(a - 2)$

15) $h(x) = 3x + 5$, find $h(6x)$

16) $h(x) = x^2 + 1$, find $h(\frac{x}{4})$

Adding and Subtracting Functions

✎*Perform the indicated operation.*

1) $h(t) = 2t + 1$

    $g(t) = 2t + 2$

    Find $(h - g)(t)$

2) $g(a) = -3^a - 3$

    $f(a) = a^2 + 5$

    Find $(g - f)(a)$

3) $g(x) = 2x - 5$

    $h(x) = 4x + 5$

    Find $g(3) - h(3)$

4) $h(3) = 3x + 3$

    $g(x) = -4x + 1$

    Find $(h + g)(10)$

5) $f(x) = 4x - 3$

    $g(x) = x^3 + 2x$

    Find $(f - g)(4)$

6) $h(n) = 4n + 5$

    $g(n) = 3n + 4$

    Find $(h - g)(n)$

7) $g(x) = -x^2 - 1 - 2x$

    $f(x) = 5 + x$

    Find $(g - f)(x)$

8) $g(t) = 2t + 5$

    $f(t) = -t^2 + 5$

    Find $(g + f)(t)$

Multiplying and Dividing Functions

✎*Perform the indicated operation.*

1) $g(a) = 2a - 1$
   $h(a) = 3a - 3$
   Find $(g.h)(-4)$

2) $f(x) = 2x^3 - 5x^2$
   $g(x) = 2x - 1$
   Find $(f.g)(x)$

3) $g(t) = t^2 + 3$
   $h(t) = 4t - 3$
   Find $(g.h)(-1)$

4) $g(n) = n^2 + 4 + 2n$
   $h(n) = -3n + 2$
   Find $(g.h)(1)$

5) $g(a) = 3a + 2$
   $f(a) = 2a - 4$
   Find $(\frac{g}{f})(3)$

6) $f(x) = 3x - 1$
   $g(x) = x^2 - x$
   Find $(\frac{f}{g})(x)$

7) $h(a) = 3a$
   $g(a) = -a^3 - 3$
   Find $(\frac{h}{g})(a)$

Composition of Functions

🖎 Using f(x) = 5x + 4 and g(x) = x − 3, find:

1) $f(g(6))$

3) $g(f(-7))$

2) $f(f(8))$

4) $g(f(x))$

🖎 Using f(x) = 6x + 2 and g(x) = x − 5, find:

5) $f(g(7))$

7) $g(f(3))$

6) $f(f(2))$

8) $g(g(x))$

🖎 Using f(x) = 7x + 4 and g(x) = 2x − 4, find:

9) $f(g(3))$

11) $g(f(4))$

10) $f(f(3))$

12) $g(g(5))$

Answers of Worksheets – Chapter 19

Function Notation

1) $d(t) = 22t$
2) $c(p) = p^2 + 5p + 5$
3) $m(n) = 25n - 120$
4) $f(x) = 2x - 6$
5) 13
6) 15

7) 0
8) −20
9) 16
10) 57
11) 12
12) −8

13) $-22a + 5$
14) $7a - 11$
15) $18x + 5$
16) $1 + \dfrac{1}{16}\, x^2$

Adding and Subtracting Functions

1) −1
2) $-a^2 - 3a - 8$
3) −16

4) −6
5) −59
6) $n + 1$

7) $-x^2 - 3x - 6$
8) $-t^2 + 2t + 10$

Multiplying and Dividing Functions

1) 135
2) $4x^4 - 12x^3 + 5x^2$
3) −28

4) −7
5) $\dfrac{11}{2}$

6) $\dfrac{3x-1}{x^2-x}$
7) $\dfrac{3a}{-a^2-3}$

Composition of functions

1) 19
2) 224
3) −34
4) $5x + 1$

5) 14
6) 86
7) 15
8) $x - 10$

9) 18
10) 179
11) 60
12) 8

Chapter 20: Probability

Math Topics that you'll learn today:

- ✓ Probability of Simple Events
- ✓ Factorials
- ✓ Permutations
- ✓ Combination

Mathematics is the supreme judge; from its decisions there is no appeal.

–Tobias Dantzig

Probability of Simple Events

✎ *Solve.*

1) A number is chosen at random from 1 to 10. Find the probability of selecting a 4 or smaller.

2) A number is chosen at random from 1 to 50. Find the probability of selecting multiples of 10.

3) A number is chosen at random from 1 to 10. Find the probability of selecting of 4 and factors of 6.

4) A number is chosen at random from 1 to 10. Find the probability of selecting a multiple of 3.

5) A number is chosen at random from 1 to 50. Find the probability of selecting prime numbers.

6) A number is chosen at random from 1 to 25. Find the probability of not selecting a composite number.

Factorials

✍*Determine the value for each expression.*

1) $\dfrac{9!}{6!}$

2) $\dfrac{8!}{5!}$

3) $\dfrac{7!}{5!}$

4) $\dfrac{20!}{18!}$

5) $\dfrac{22!}{18!5!}$

6) $\dfrac{10!}{8!2!}$

7) $\dfrac{100!}{97!}$

8) $\dfrac{14!}{10!4!}$

9) $\dfrac{10!}{8!}$

10) $\dfrac{25!}{20!}$

11) $\dfrac{14!}{9!3!}$

12) $\dfrac{55!}{53!}$

13) $\dfrac{(2.3)!}{3!}$

14) $5! + 4!$

Permutations

✍ *Evaluate each expression.*

1) $_4P_2$

2) $_5P_1$

3) $_6P_2$

4) $_6P_6$

5) $-4 + _7P_4$

6) $5 \cdot _6P_5$

7) $_7P_2$

8) $_4P_1$

9) $_8P_5$

10) $_7P_3$

11) How many possible 7–digit telephone numbers are there? Someone left their umbrella on the subway and we need to track them down.

12) With repetition allowed, how many ways can one choose 8 out of 12 things?

Combination

✍ List all possible combinations.

1) 4, 5, 6, 7, taken four at a time

2) T, V, W, taken two at a time

✍ Evaluate each expression.

3) $_7C_5$

4) $_4C_2$

5) $_9C_3$

6) $_5C_2$

7) $_{12}C_8$

8) $_9C_6$

9) $_{22}C_{20}$

10) $_{12}C_8$

11) $_{11}C_8$

12) $_{25}C_{23}$

13) $_{17}C_{10}$

14) $_{24}C_5$

15) $4 \cdot {}_{18}C_{11}$

16) $_{20}C_{16} + 1$

Answers of Worksheets – Chapter 20

Probability of simple events

1) $\frac{2}{5}$ 3) $\frac{1}{5}$ 5) $\frac{3}{10}$

2) $\frac{1}{10}$ 4) $\frac{3}{10}$ **6)** $\frac{2}{5}$

Factorials

1) 504   6) 45   11) 40,040

2) 336   7) 970,200   12) 2,970

3) 42   8) 1,001   13) 120

4) 380   9) 90   14) 144

5) 1,463   10) 6,375,600

Permutations

1) 12   5) 836   9) 6,720

2) 5   6) 3, 600   10) 210

3) 30   7) 42   13) 10^7

4) 720   8) 4   14) 12^8

Combination

1) 4567   7) 495   13) 19,448

2) TV VW TW   8) 84   14) 42,504

3) 27   9) 231   15) 127,296

4) 6   10) 495   16) 4, 846

5) 84   11) 165   17) 11,622

6) 10   12) 300

Chapter 21: Trigonometric Functions

Math Topics that you'll learn today:

- ✓ Right Triangle Trigonometry
- ✓ Sketch Each Angle in Standard Position
- ✓ Finding Co–Terminal Angles and Reference Angles
- ✓ Finding the Cosine and Sine of Each Angle
- ✓ Writing Each Measure in Radians
- ✓ Writing Each Measure in Degrees
- ✓ Evaluating Each Trigonometric Expression
- ✓ Missing Sides and Angles of a Right Triangle
- ✓ Arc Length and Sector Area
- ✓ Trig Ratios of General Angles

Mathematics is like checkers in being suitable for the young, not too difficult, amusing, and without peril

to the state. — Plato

Right Triangle Trigonometry

✍ **Find the value of the trig function indicated.**

1) Find cot θ if cos θ = $\frac{3}{13}$

2) Find cot θ if cos θ = $\frac{\sqrt{2}}{10}$

3) Find cot θ if cos θ = $\frac{3}{5}$

4) Find cot θ if cos θ = $\frac{\sqrt{10}}{10}$

5) Find cot θ if cos θ = $\frac{4\sqrt{13}}{17}$

6) Find cot θ if cos θ = $\frac{4}{5}$

✍ **Find the measure of each side indicated. Round to the nearest tenth.**

7)

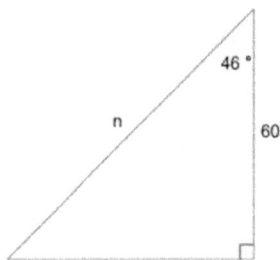

8)

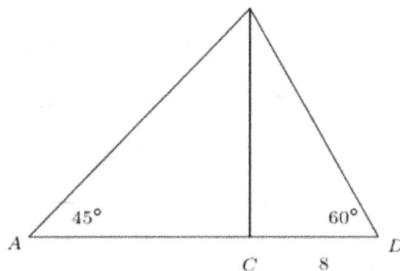

9)

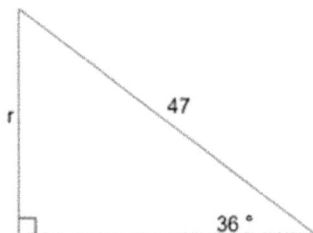

10)

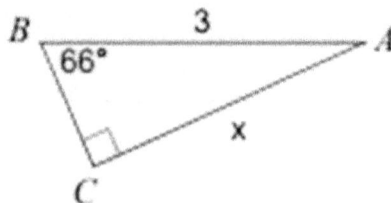

Sketch Each Angle in Standard Position

✍️*Draw the angle with the given measure in standard position.*

1) −120°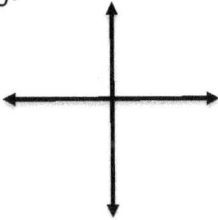

4) 280°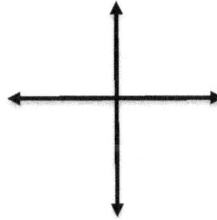

2) 440°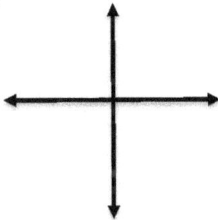

5) 710°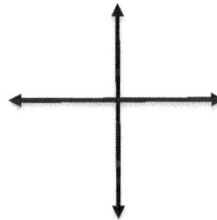

3) $-\frac{10\pi}{3}$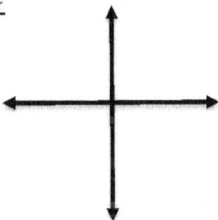

6) $\frac{11\pi}{6}$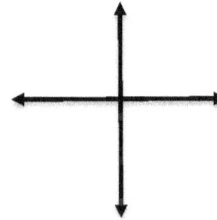

Finding Co–Terminal Angles and Reference Angles

✍️ *Find a conterminal angle between 0° and 360°.*

1) $-440°$

2) $640°$

3) $-435°$

4) $-330°$

✍️ *Find a conterminal angle between 0 and 2π for each given angle.*

5) $\dfrac{15\pi}{4}$

6) $-\dfrac{19}{12}$

7) $-\dfrac{35\pi}{18}$

8) $\dfrac{11}{3}$

✍️ *Find the reference angle.*

9)

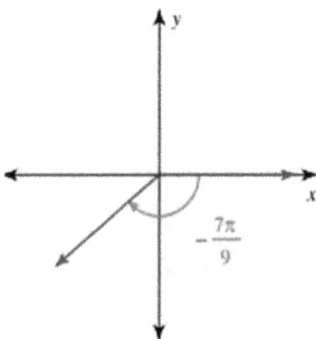

10)

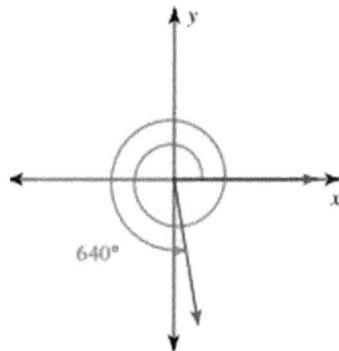

Finding the Cosine and Sine of Each Angle

✍ **Find each measurement indicated. Round your answers to the nearest tenth.**

1) Find AB

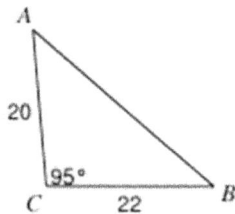

3) Find AB

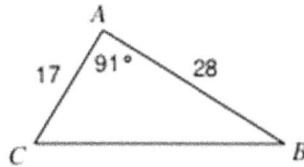

2) Find AB

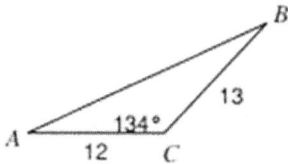

4) Find AB

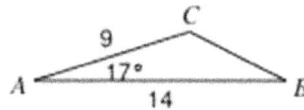

✍ **Find each angle measure to the nearest degree.**

5) tan C = 0.1405

6) cos Y = 0.5736

✍ **Find the value of each trigonometric ratio to the nearest ten– thousandth.**

7) sin 14∘

9) tan 79∘

8) cos 31∘

10) cos 60

Writing Each Measure in Radians

✍️*Convert each degree measure into radians.*

1) −140°

2) 320°

3) 210°

4) 970°

5) −190°

6) 345°

7) 265°

8) 555°

9) 300°

10) 50°

11) 315°

12) 600°

13) 712°

14) −160°

15) −210°

16) 545°

17) −30°

18) 660°

19) −170°

20) 230°

21) 150°

Writing Each Measure in Degrees

✎ Convert each radian measure into degrees.

1) $\frac{\pi}{30}$

2) $\frac{32\pi}{40}$

3) $\frac{14\pi}{36}$

4) $\frac{\pi}{5}$

5) $-\frac{10\pi}{8}$

6) $\frac{14\pi}{3}$

7) $-\frac{16\pi}{3}$

8) $-\frac{50\pi}{14}$

9) $\frac{11\pi}{6}$

10) $\frac{5\pi}{9}$

11) $-\frac{\pi}{3}$

12) $\frac{13\pi}{6}$

13) $\frac{15\pi}{20}$

14) $\frac{21\pi}{4}$

15) $-\frac{68\pi}{45}$

16) $\frac{14\pi}{3}$

17) $-\frac{41\pi}{12}$

18) $-\frac{17\pi}{9}$

19) $\frac{35}{18}$

20) $-\frac{3\pi}{2}$

21) $\frac{4\pi}{9}$

Evaluating Each Trigonometric

✎ *Find the exact value of each trigonometric function.*

1) $\cos 225°$

2) $\tan \dfrac{7\pi}{6}$

3) $\tan -\dfrac{\pi}{6}$

4) $\cot -\dfrac{7\pi}{6}$

5) $\cos -\dfrac{\pi}{4}$

6) $\cos -480°$

7) $\sin 690°$

8) $\tan 420°$

9) $\cot -495°$

10) $\tan 405°$

✎ *Use the given point on the terminal side of angle θ to find the value of the trigonometric function indicated.*

11) $\sin \theta;\ (-6, 4)$

12) $\cos \theta;\ (2, -2)$

13) $\cot \theta;\ (-7, \sqrt{15})$

14) $\cos \theta;\ (-2\sqrt{3}, -2)$

15) $\sin \theta;\ (-\sqrt{7}, 3)$

16) $\tan \theta;\ (-11, -2)$

Missing Sides and Angles of a Right Triangle

🖎*Find the value of each trigonometric ratio as fractions in their simplest form.*

1) tan A

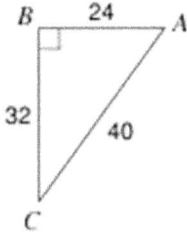

2) sin X

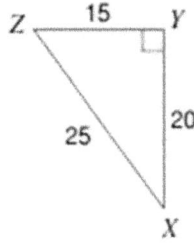

🖎*Find the missing side. Round answers to the nearest tenth.*

3)

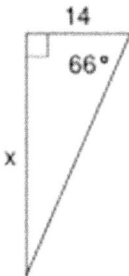

4)

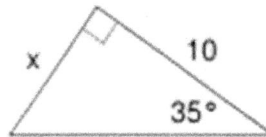

5)

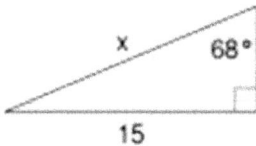

6)

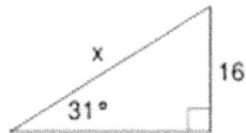

Arc Length and Sector Area

✎ **Find the length of each arc. Round your answers to the nearest tenth.**

1) r = 28 cm, θ = 45∘

3) r = 22 ft, θ = 60∘

2) r = 15 ft, θ = 95∘

4) r = 12 m, θ = 85∘

✎ **Find area of a sector. Do not round.**

5)

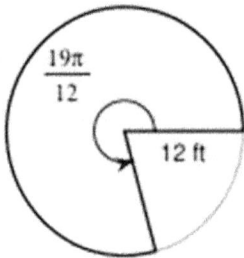

7)

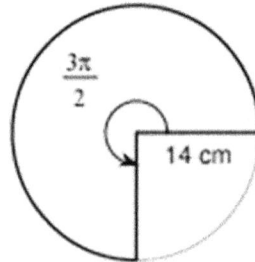

6)

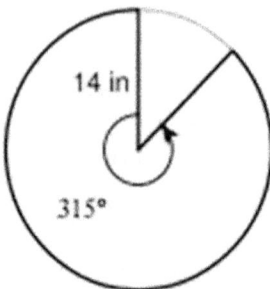

8)

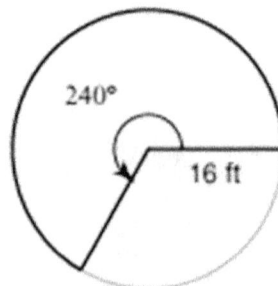

Trig Ratios of General Angles

✏️*Use a calculator to find each. Round your answers to the nearest ten–thousandth.*

1) $\sin - 120°$

2) $\sin - 228°$

3) $\cos 310°$

4) $\cos 101°$

5) $\sin 105°$

6) $\sin - 305°$

✏️*Find the exact value of each trigonometric function. Some may be undefined.*

7) $\sec 0$

8) $\tan - \dfrac{3\pi}{2}$

9) $\cos \dfrac{11\pi}{6}$

10) $\cot \dfrac{5\pi}{3}$

11) $\sec - \dfrac{3\pi}{4}$

12) $\tan \dfrac{2\pi}{3}$

Answers of Worksheets – Chapter 21

Right triangle trigonometry

1) $\frac{3\sqrt{10}}{40}$

2) $\frac{1}{7}$

3) $\frac{3}{4}$

4) $\frac{1}{3}$

5) $\frac{4\sqrt{13}}{9}$

6) $\frac{4}{3}$

7) 86.4

8) $8\sqrt{3}$

9) 27.6

10) 2.7

Sketch each angle in standard position

1) −120∘

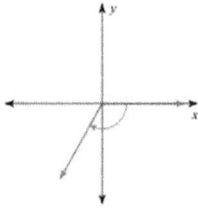

4) 280∘

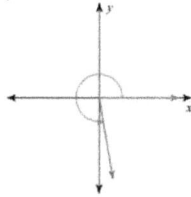

2) 440∘

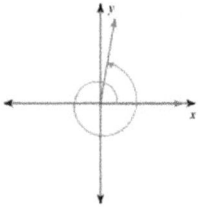

5) 710∘

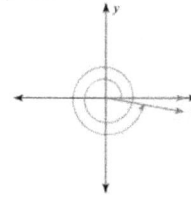

3) $-\frac{10\pi}{3}$

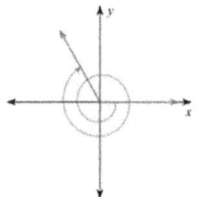

6) $\frac{11\pi}{6}$

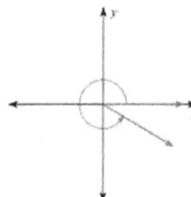

Finding co–terminal angles and reference angles

1) 280∘	5) $\frac{7\pi}{4}$	8) $\frac{5\pi}{3}$
2) 280∘	6) $\frac{5\pi}{12}$	9) $\frac{2\pi}{9}$
3) 285∘	7) $\frac{\pi}{18}$	10) 80∘
4) 30∘		

Finding the cosine and sine of each angle

1) 31	5) 8∘	9) 5.1446
2) 23	6) 55∘	10) 0.5
3) 33	7) 0.2419	
4) 6	8) 0.8572	

Writing each measure in radians

1) $-\frac{7\pi}{9}$	8) $\frac{37}{12}$	15) $-\frac{7\pi}{6}$
2) $\frac{16}{9}$	9) $\frac{5\pi}{3}$	16) $\frac{109\pi}{36}$
3) $\frac{7\pi}{6}$	10) $\frac{5\pi}{18}$	17) $-\frac{\pi}{6}$
4) $\frac{97\pi}{18}$	11) $\frac{7\pi}{4}$	18) $\frac{11}{3}$
5) $-\frac{19}{18}$	12) $\frac{10\pi}{3}$	19) $-\frac{17\pi}{18}$
6) $\frac{23}{12}$	13) $\frac{178\pi}{45}$	20) $\frac{23\pi}{18}$
7) $\frac{53}{36}$	14) $-\frac{8\pi}{9}$	21) $\frac{5\pi}{6}$

Writing each measure in degrees

1) 6∘	5) −225∘	9) 330∘
2) 144∘	6) 840∘	10) 100∘
3) 70∘	7) −960∘	11) −60∘
4) 36∘	8) −643∘	12) 390∘

13) 135∘   16) 840∘   19) 350∘

14) 945∘   17) –615∘   20) –270∘

15) –272∘   18) –340∘   21) 80∘

Evaluating each trigonometric expression

1) $-\frac{\sqrt{2}}{2}$ 6) $-\frac{1}{2}$ 12) $-\sqrt{2}$

2) $\frac{\sqrt{3}}{3}$ 7) $-\frac{1}{2}$ 13) $-\frac{7\sqrt{15}}{15}$

3) $-\frac{\sqrt{3}}{3}$ 8) $\sqrt{3}$ 14) $-\frac{\sqrt{3}}{2}$

                        9) 1   15) $\frac{3}{4}$

4) $-\sqrt{3}$

5) $\frac{\sqrt{2}}{2}$ 10) 1   16) $\frac{2}{11}$

                        11) $\frac{2\sqrt{13}}{13}$

Missing sides and angles of a right triangle

1) $\frac{4}{3}$ 3) 31.4   6) 31.1

                        4) 7.0

2) $\frac{3}{5}$ 5) 16.2

Arc length and sector area

1) 22 cm   5) 114π ft²   8) $\frac{512\pi}{3}$ ft²

2) 25 ft   6) $\frac{343\pi}{2}$ in²

3) 23 ft   7) 147π cm²

4) 18 m

Trig ratios of general angles

1) −0.8660   6) 0.8192   10) $-\frac{\sqrt{3}}{3}$

2) 0.7431   7) 1

3) 0.6428   8) Undefined   11) $-\sqrt{2}$

4) −0.1908   9) $\frac{\sqrt{3}}{2}$ 12) $-\sqrt{3}$

5) 0.9659

Chapter 22: Conic Sections

Math Topics that you'll learn today:

- ✓ Parabolas: Write The Equation of The Parabola in Standard Form
- ✓ Finding The Focus, Vertex, and The Directrix of The Parabola
- ✓ Writing The Equation for The Parabola
- ✓ Writing The Standard Form of The Circle
- ✓ Finding The Center and The Radius of Circles
- ✓ Writing The Standard Equation of Each Ellipse
- ✓ Finding the Foci, Vertices, and Co– Vertices of Ellipses
- ✓ Finding The Vertices, Co– Vertices, Foci, and Asymptotes of The Hyperbola
- ✓ Writing The Equation of The Hyperbola in Standard Form
- ✓ Classifying a Conic Section (in Standard Form)
- ✓ Classifying a Conic Section (Not in Standard Form)

Mathematics is as much an aspect of culture as it is a collection of algorithms. — Carl Boyer

Parabolas: Write The Equation of The Parabola in Standard Form

✍ *Write the equation of the following parabolas.*

1) Vertex (0, 0) and Focus (0, 2)

2) Vertex (3, 2) and Focus (3, 4)

3) Vertex (1, 1) and Focus (1, 6)

4) Vertex (− 1, 2) and Focus (− 1, 5)

5) Vertex (2, 2) and Focus (2, 6)

6) Vertex (0, 1) and Focus (0, 2)

7) Vertex (2, 1) and Focus (4, 1)

8) Vertex (5, 0) and Focus (9, 0)

9) Vertex (− 2, 4) and Focus (2, 4)

10) Vertex (− 4, 2) and Focus (0, 2)

216

Finding The Focus, Vertex, and The Directrix of The Parabola

✎Use the information provided to write the vertex form equation of each parabola.

1) $y = x^2 + 8x$

2) $y = x^2 - 6x + 5$

3) $y + 6 = (x + 3)^2$

4) $y = x^2 + 10x + 33$

5) $y = (x + 5)(x + 4)$

6) $\frac{1}{2}(y + 4) = (x - 7)^2$

7) $162 + 731 = -y - 9x^2$

8) $y = x^2 + 16x + 71$

9) Focus: $(-\frac{63}{8}, -7)$, Directrix: $x = -\frac{65}{8}$

10) Focus: $(\frac{107}{12}, -7)$, Directrix: $x = \frac{109}{12}$

11) Opens up or down, and passes through (–6, –7), (–11, –2), and (–8, 1)

12) Opens up or down, and passes through (11, 15), (7, 7), and (4, 22)

Writing The Equation for The Parabola

✍ *Use the information provided to write the vertex form equation of each parabola.*

1) Vertex: $(8, 9)$, Directrix: $y = \dfrac{73}{8}$

2) Vertex: $(-5, 8)$, Focus: $(-\dfrac{21}{4}, 8)$

3) Vertex: $(-6, -9)$, Directrix: $x = -\dfrac{47}{8}$

4) Vertex: $(5, -1)$, y–intercept: $-\dfrac{27}{2}$

5) Vertex: $(8, -1)$, y–intercept: -17

6) Vertex at origin, Directrix: $y = -\dfrac{1}{8}$

7) Vertex at origin, Directrix: $y = \dfrac{1}{4}$

8) Vertex at origin, Focus: $(0, -\dfrac{1}{32})$

9) Opens left or right, Vertex: $(7, 0)$, Passes through: $(6, -1)$

10) Opens left or right, Vertex: $(7, 6)$, Passes through: $(-11, 9)$

11) Focus: $(-\dfrac{63}{8}, -7)$, Directrix: $x = -\dfrac{65}{8}$

12) Focus: $(\dfrac{107}{12}, -7)$, Directrix: $x = \dfrac{109}{12}$

Writing The Standard Form of The Circle

✍ *Use the information provided to write the standard form equation of each circle.*

1) $x^2 + y^2 - 8x - 6y + 21 = 0$

2) $y^2 + 2x + x^2 = 24y - 120$

3) $x^2 + y^2 - 2y - 15 = 0$

4) $8x + x^2 - 2y = 64 - y^2$

5) Center: (−5, −6), Radius: 9

6) Center: (−9, −12), Radius: 4

7) Center: (−12, −5), Area: 4π

8) Center: (−11, −14), Area: 16π

9) Center: (−3, 2), Circumference: 2π

10) Center: (15, 14), Circumference: $2\pi\sqrt{15}$

Finding The Center and The Radius of Circles

✍ *Identify the center and radius of each. Then sketch the graph.*

1) $(x - 2)^2 + (y + 5)^2 = 10$

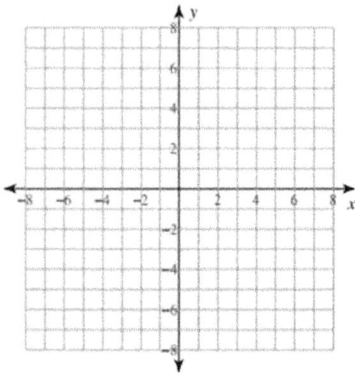

2) $x^2 + (y - 1)^2 = 4$

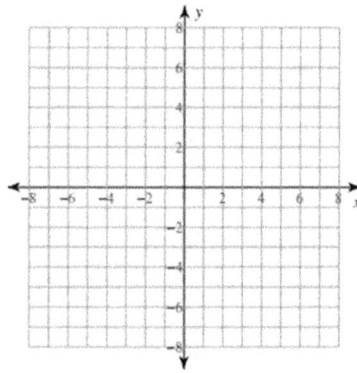

3) $(x - 2)^2 + (y + 6)^2 = 9$

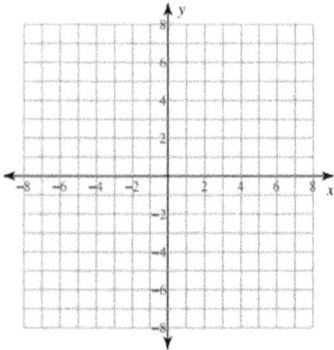

4) $(x + 14)^2 + (y - 5)^2 = 16$

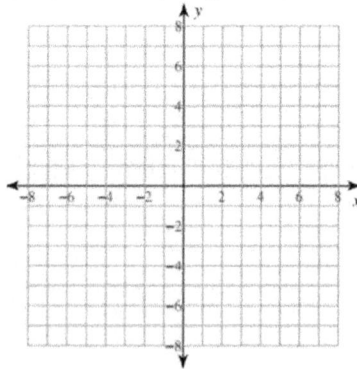

Writing the standard equation of each ellipse

✎*Use the information provided to write the standard form equation of each ellips.*

1) Foci: $(2\sqrt{3}, 0)$, $(-2\sqrt{3}, 0)$
   Co–vertices: $(0, 2)$. $(0, -2)$

2) Vertices: $(0, 6)$, $(0, -6)$
   Co–vertices: $(3, 0)$. $(-3, 0)$

3) Vertices: $(4, 3)$, $(4, -7)$
   Co–vertices: $(1, -2)$. $(7, -2)$

4) Foci: $(\sqrt{17}, 0)$, $(-\sqrt{17}, 0)$
   Co–vertices: $(9, 0)$. $(-9, 0)$

5) Foci: $(-7, 5 + \sqrt{13})$, $(-7, 5 - \sqrt{13})$
   Co–vertices: $(-1, 5)$. $(-13, 5)$

6) Vertices: $(5, 1)$, $(-1, 1)$
   Co–vertices: $(2, 3)$. $(2, -1)$

7) Vertices: $(12, 0)$, $(-12, 0)$
   Co–vertices: $(2\sqrt{11}, 0)$. $(-2\sqrt{11}, 0)$

8) Vertices: $(7 + 2\sqrt{35}, -4)$, $(7 - 2\sqrt{35}, -4)$
   Co–vertices: $(7, -2)$. $(7, -6)$

9) Center: $(4, 8)$
   Vertex: $(4, 8 - \sqrt{170})$
   Co–vertex: $(4 - \sqrt{15}, 8)$

10) Center: $(7, -10)$
    Vertex: $(-6, -10)$
    Co–vertex: $(7, -17)$

Finding The Foci, Vertices, and Co– Vertices of Ellipses

✎ *Identify the vertices, co–vertices, foci.*

1) $\dfrac{x^2}{169} + \dfrac{y^2}{64} = 1$

2) $\dfrac{x^2}{95} + \dfrac{y^2}{30} = 1$

3) $\dfrac{x^2}{36} + \dfrac{y^2}{16} = 1$

4) $\dfrac{x^2}{49} + \dfrac{y^2}{169} = 1$

5) $\dfrac{(x+5)^2}{81} + \dfrac{(y-1)^2}{144} = 1$

6) $\dfrac{(x-3)^2}{49} + \dfrac{(y-9)^2}{4} = 1$

7) $\dfrac{x^2}{64} + \dfrac{(y-8)^2}{9} = 1$

8) $\dfrac{x^2}{64} + \dfrac{(y-6)^2}{121} = 1$

Graph each equation.

9) $\dfrac{x^2}{36} + \dfrac{y^2}{25} = 1$

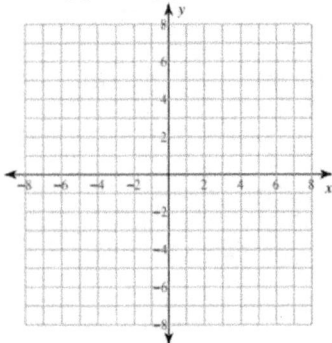

10) $\dfrac{x^2}{9} + \dfrac{y^2}{49} = 1$

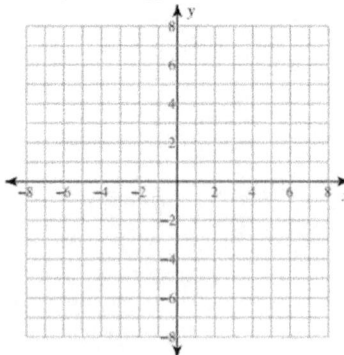

Finding The Vertices, Co– Vertices, Foci, and Asymptotes of The Hyperbola

✍ *Identify the vertices, foci, and direction of opening of each.*

1) $\dfrac{y^2}{25} - \dfrac{x^2}{16} = 1$

4) $\dfrac{x^2}{81} - \dfrac{y^2}{4} = 1$

2) $\dfrac{x^2}{121} - \dfrac{y^2}{36} = 1$

5) $\dfrac{(x+2)^2}{169} - \dfrac{(y+8)^2}{4} = 1$

3) $\dfrac{x^2}{121} - \dfrac{y^2}{81} = 1$

6) $\dfrac{(y+8)^2}{36} - \dfrac{(y+2)^2}{25} = 1$

✍ *Identify the vertices and foci of each. Then sketch the graph.*

7) $\dfrac{y^2}{25} - \dfrac{x^2}{25} = 1$

8) $\dfrac{x^2}{25} - \dfrac{(y-2)^2}{4} = 1$

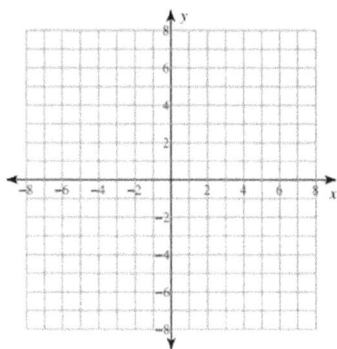

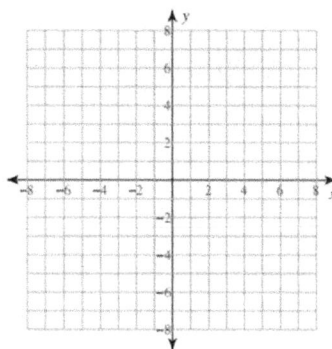

Writing The Equation of The Hyperbola in Standard Form

✍ *Use the information provided to write the standard form equation of each hyperbola.*

1) $-2x^2 + 3y^2 + 4x - 60y + 268 = 0$

2) $-x^2 + y^2 - 18x - 14y - 132 = 0$

3) $-16x^2 + 9y^2 + 32x + 144y - 16 = 0$

4) $9x^2 - 4y^2 - 90x + 32y - 163 = 0$

5) Vertices: (8, 14), (8, −10), Conjugate Axis is 6 units long

6) Vertices: (7, 4), (7, −24), Distance from Center to Focus = $7\sqrt{5}$

7) Vertices: (−5, 22), (−5, −4), Distance from Center to Focus = $\sqrt{218}$

8) Vertices: (0, −1), (−20, −1), Asymptotes: $y = x + 9$, $y = -x - 11$

9) Foci: (−9, −5 + $9\sqrt{2}$), (−9, −5 − $9\sqrt{2}$), Conjugate Axis is 18 units long

10) Foci: (8, −5 + $\sqrt{53}$), (8, −5 −$\sqrt{53}$),

   Endpoints of Conjugate Axis: (15, −5), (1, −5)

Classifying a Conic Section (in Standard Form)

✍*Classify each conic section and write its equation in standard form.*

1) $x^2 - 4y^2 + 6x - 8y + 1 = 0$

2) $3x^2 + 3x + y + 79 = 0$

3) $x^2 + y^2 + 4x - 2y - 18 = 0$

4) $-y^2 + x + 8y - 17 = 0$

5) $49x^2 + 9y^2 + 392x + 343 = 0$

6) $-9x^2 + y^2 - 72x - 153 = 0$

7) $-2y^2 + x - 20y - 49 = 0$

8) $-x^2 + 10x + y - 21 = 0$

✍*Classify each conic section, erite its equation in standard form, and sketch its graph. For parabolas, identify the vertex and focus. For circles, identify the center and radius. For ellises and hyperbolas identify the center, vertices, and foci.*

9) $9x^2 + 4y^2 - 54x - 8y - 59 = 0$

10) $y^2 + x + 10y + 26 = 0$

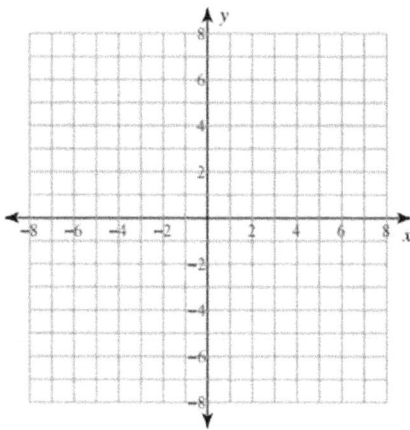

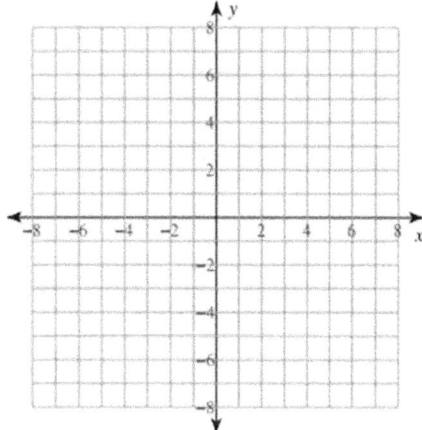

Classifying a Conic Section (Not in Standard Form)

✍ *Classify each conic section.*

1) $x^2 + y^2 - 8x + 8y - 4 = 0$

2) $y = 6x^2 - 60x + 149$

3) $x^2 - 4x + 4y^2 - 32y + 32 = 0$

4) $x^2 - 2x - 36y^2 - 360y - 935 = 0$

5) $y = 6x^2 - 60x + 149$

6) $x^2 + y^2 - 8x + 8y - 4 = 0$

7) $x^2 + y^2 + 6x + 10y + 33 = 0$

8) $x^2 - 4x - 36y^2 + 288y - 608 = 0$

9) $9x^2 + 4y^2 + 16y - 128 = 0$

10) $x^2 + 8x - 25y^2 + 50y - 34 = 0$

11) $y = 6x^2 + 60x + 155$

12) $4x^2 + 9y^2 - 54y + 45 = 0$

13) $-9x^2 - 54x + 4y^2 - 40y - 125 = 0$

14) $x^2 - 4x + 4y^2 - 32y + 32 = 0$

Answers of Worksheets – Chapter 22

Parabolas: Write the equation of the parabola in standard form

1) Write the equation of the following parabolas.

2) Vertex (0, 0) and Focus (0, 2): $x^2 = 8y$

3) Vertex (3, 2) and Focus (3, 4): $(x - 3)^2 = 8\ (y - 2)$

4) Vertex (1, 1) and Focus (1, 6): $(x - 1)^2 = 20\ (y - 1)$

5) Vertex (− 1, 2) and Focus (− 1, 5): $(x + 1)^2 = 12\ (y - 2)$

6) Vertex (2, 2) and Focus (2, 6): $(x - 2)^2 = 8\ (y - 2)$

7) Vertex (0, 1) and Focus (0, 2): $x^2 = 8\ (y - 1)$

8) Vertex (2, 1) and Focus (4, 1): $(y - 1)^2 = 8\ (x - 2)$

9) Vertex (5, 0) and Focus (9, 0): $(y - 1)^2 = 8\ (x - 2)$

10) Vertex (− 2, 4) and Focus (2, 4): $(y - 4)^2 = 16\ (x + 2)$

11) Vertex (− 4, 2) and Focus (0, 2): $(y + 4)^2 = 16x$

Finding the focus, vertex, and the directrix of the Parabola

12) $y = (x + 4)^2 - 16$

13) $y = (x - 3)^2 - 4$

14) $y = (x + 3)^2 - 6$

15) $y = (x + 5)^2 + 8$

16) $y = (x + \frac{9}{2})^2 - \frac{1}{4}$

17) $y = 2\ (x - 7)^2 - 4$

18) $y = -9\ (x + 9)^2 - 2$

19) $y = (x + 8)^2 + 7$

20) $x = 2\ (y + 7)^2 - 8$

21) $x = -3\ (y + 7)^2 + 9$

22) $y = -\ (x + 9)^2 + 2$

23) $y = (x - 8)^2 + 6$

Writing the equation for the parabola

1) $y = -2\ (x - 8)^2 + 9$

2) $x = -(y - 8)^2 - 5$

3) $x = -2\ (y + 9)^2 - 6$

4) $y = -\frac{1}{2}\ (x - 5)^2 - 1$

5) $y = -\frac{1}{4}\ (x - 8)^2 - 1$

6) $y = 2x^2$

7) $y = -x^2$

8) $y = -8x^2$

9) $x = -y^2 + 7$

10) $x = -2\,(y - 6)^2 + 7$

11) $x = 2\,(y + 7)^2 - 8$

12) $x = -3\,(y + 7)^2 + 9$

Writing the standard form of the circle

1) $(x - 4)^2 + (y - 3)^2 = 4$

2) $(x + 1)^2 + (y - 12)^2 = 25$

3) $x^2 + (y - 1)^2 = 16$

4) $(x + 4)^2 + (y - 1)^2 = 81$

5) $(x + 5)^2 + (y + 6)^2 = 81$

6) $(x + 9)^2 + (y + 12)^2 = 16$

7) $(x + 12)^2 + (y + 5)^2 = 4$

8) $(x + 11)^2 + (y + 14)^2 = 16$

9) $(x + 3)^2 + (y - 2)^2 = 1$

10) $(x - 15)^2 + (y - 14)^2 = 15$

Finding the center and the radius of circles

1) Center: (2, –5), Radius: $\sqrt{10}$

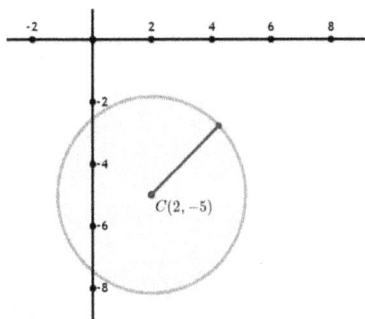

2) Center: (0, 1), Radius: $2\sqrt{26}$

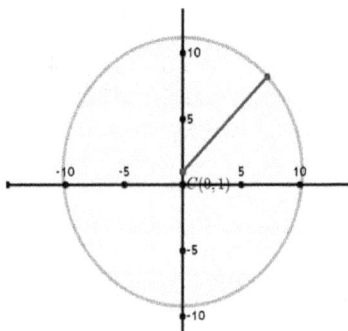

3) Center: $(2, -6)$, Radius: 3

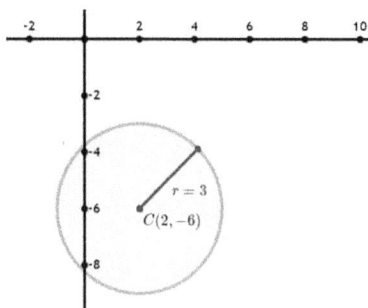

4) Center: $(-14, -5)$, Radius: 4

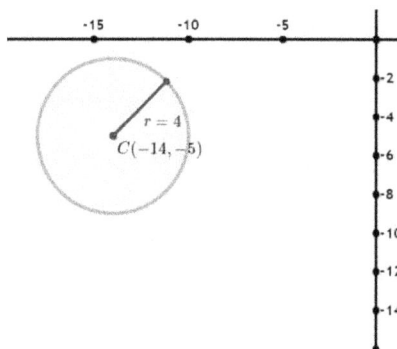

Writing the standard equation of each ellipse

1) $\dfrac{x^2}{16} + \dfrac{y^2}{4} = 1$

2) $\dfrac{x^2}{9} + \dfrac{y^2}{36} = 1$

3) $\dfrac{(x-4)^2}{9} + \dfrac{(y+2)^2}{25} = 1$

4) $\dfrac{x^2}{81} + \dfrac{y^2}{64} = 1$

5) $\dfrac{(x+7)^2}{36} + \dfrac{(y-5)^2}{49} = 1$

6) $\dfrac{(x-2)^2}{9} + \dfrac{(y-1)^2}{4} = 1$

7) $\dfrac{x^2}{144} + \dfrac{y^2}{100} = 1$

8) $\dfrac{(x-7)^2}{144} + \dfrac{(y+5)^2}{4} = 1$

9) $\dfrac{(x-4)^2}{15} + \dfrac{(y-8)^2}{170} = 1$

10) $\dfrac{(x-7)^2}{169} + \dfrac{(y+10)^2}{49} = 1$

Finding the foci, vertices, and co–vertices of ellipses

1) Vertices: $(13, 0)$, $(-13, 0)$

Co–vertices: $(0, 8)$, $(0, -8)$

Foci: $(\sqrt{105}, 0)$, $(-\sqrt{105}, 0)$

2) Vertices: $(\sqrt{95}, 0)$, $(-\sqrt{95}, 0)$

Co–vertices: $(0, \sqrt{30})$, $(0, -\sqrt{30})$

Foci: $(\sqrt{65}, 0)$, $(-\sqrt{65}, 0)$

3) Vertices: (6, 0), (−6, 0)

Co−vertices: (0,4), (0, −4)

Foci: $(2\sqrt{5}, 0), (−2\sqrt{5}, 0)$

4) Vertices: (0, 13), (0, −13)

Co−vertices: (7, 0), (−7, 0)

Foci: $(0, 2\sqrt{30}), (0, −2\sqrt{30})$

5) Vertices: (−5, 13), (−5, −11)

Co−vertices: (4, 1), (−14, 1)

Foci: $(−5, 1 + 3\sqrt{7}), (−5, 1 − 3\sqrt{7})$

6) Vertices: (10, 9), (−4, 9)

Co−vertices: (3, 11), (3, 7)

Foci: $(3 + 3\sqrt{5}, 9), (3 − 3\sqrt{5}, 9)$

7) Vertices: (8, 8), (−8, 8)

Co−vertices: (0, 11), (0, 5)

Foci: $(\sqrt{55}, 8), (−\sqrt{55}, 8)$

8) Vertices: (0, 17), (0, −5)

Co−vertices: (8, 6), (−8, 6)

Foci: $(0, 6 + \sqrt{57}), (0, 6 − \sqrt{57})$

11) $\dfrac{x^2}{16} + \dfrac{y^2}{36} = 1$

12) $\dfrac{x^2}{9} + \dfrac{y^2}{49} = 1$

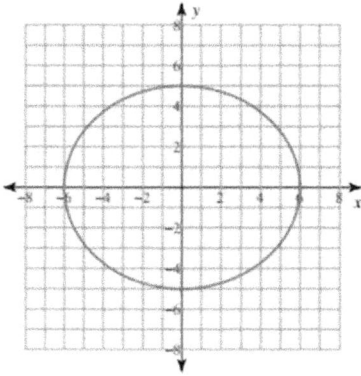

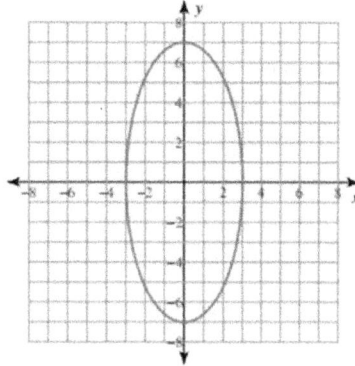

Finding the vertices, co–vertices, foci, and asymptotes of the hyperbola

1) Vertices: (0, 5), (0, −5)

Foci: $(0, \sqrt{41})$, $(0, -\sqrt{41})$

Opens up/down

2) Vertices: (11, 0), (−11, 0)

Foci: $(\sqrt{157}, 0)$, $(-\sqrt{157}, 0)$

Opens left/right

3) Vertices: (11, 0), (−11, 0)

Foci: $(\sqrt{202}, 0)$, $(-\sqrt{202}, 0)$

Opens left/right

4) Vertices: (9, 0), (−9, 0)

Foci: $(\sqrt{85}, 0)$, $(-\sqrt{85}, 0)$

Opens left/right

5) Vertices: (11, −8), (−15, −8)

Foci: $(-2 + \sqrt{173}, -8)$, $(-2 - \sqrt{173}, -8)$

Opens left/right

6) Vertices: (−2, −2), (−2, −14)

Foci: (−2, −8 + $\sqrt{61}$), (−2, −8 − $\sqrt{61}$)

Opens up/down

7) Vertices: (0, 5), (0, −5)
Foci: (0, 5$\sqrt{2}$), (0, −5$\sqrt{2}$)

8) Vertices: (5, 2), (−5, 2)
Foci: ($\sqrt{29}$, 2), (−$\sqrt{29}$, 2)

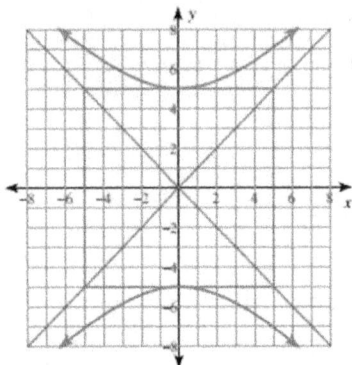

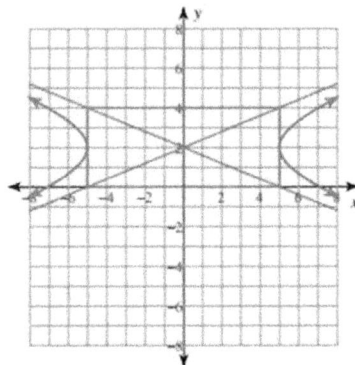

Writing the equation of the hyperbola in standard form

1) $\dfrac{(y-10)^2}{10} - \dfrac{(x-1)^2}{15} = 1$

2) $\dfrac{(y-7)^2}{100} - \dfrac{(x+9)^2}{100} = 1$

3) $\dfrac{(y+8)^2}{64} - \dfrac{(x-1)^2}{36} = 1$

4) $\dfrac{(x-5)^2}{36} - \dfrac{(y-4)^2}{81} = 1$

5) $\dfrac{(y-2)^2}{144} - \dfrac{(x-8)^2}{9} = 1$

6) $\dfrac{(y+10)^2}{196} - \dfrac{(x-7)^2}{49} = 1$

7) $\dfrac{(y-9)^2}{196} - \dfrac{(x+5)^2}{49} = 1$

8) $\dfrac{(x+10)^2}{100} - \dfrac{(y+1)^2}{100} = 1$

9) $\dfrac{(y+5)^2}{81} - \dfrac{(x+9)^2}{81} = 1$

10) $\dfrac{(y+5)^2}{4} - \dfrac{(x-8)^2}{49} = 1$

Classifying a conic section (in standard form)

1) Hyperbola, $\dfrac{(x+3)^2}{4} - (y+1)^2 = 1$

2) Parabola, $y = -3(x+5)^2 - 4$

3) Circle, $(x+2)^2 + (y-1)^2 = 23$

4) Parabola, $x = (y-4)^2 + 1$

5) Ellipse, $\dfrac{(x+4)^2}{9} + \dfrac{y^2}{49} = 1$

6) Hyperbola, $\dfrac{y^2}{9} - (x+4)^2 = 1$

7) Parabola, $x = 2(y + 5)^2 - 1$
8) Parabola, $y = (x - 5)^2 - 4$

9)
Ellips $\dfrac{(x - 3)^2}{16} - \dfrac{(y - 1)^2}{36} = 1$
Center: (3, 1)
Vertices: (3, 7), (3, −5)
Foci: $(3, 1 + 2\sqrt{5})$, $(3, 1 - 2\sqrt{5})$

10) Prabola
$x = -(y + 5)^2 - 1$
Vertex: (−1, −5)
Focus: $(-\dfrac{5}{4}, -5)$

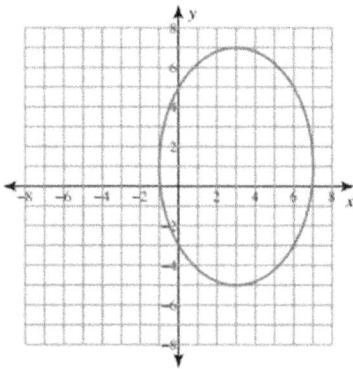

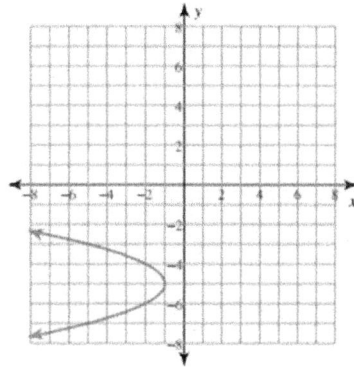

Classifying a conic section (not in standard form)

1) Circle

2) Parabola

3) Ellipse

4) Hyperbola

5) Parabola

6) Circle

7) Circle

8) Hyperbola

9) Ellipse

10) Hyperbola

11) Parabola

12) Ellipse

13) Hyperbola

14) Ellipse

Chapter 23: Sequences and Series

Math Topics that you'll learn today:

- ✓ Arithmetic Sequences
- ✓ Geometric Sequences
- ✓ Comparing Arithmetic and Geometric Sequences
- ✓ Finite Geometric Series
- ✓ Infinite Geometric Series

Mathematics is like checkers in being suitable for the young, not too difficult, amusing, and without

peril to the state. — Plato

Arithmetic Sequences

🖎*Given the first term and the common difference of an arithmetic sequence find the first five terms and the explicit formula.*

1) $a_1 = 24$, $d = 2$

2) $a_1 = -15$, $d = -5$

3) $a_1 = 18$, $d = 10$

4) $a_1 = -38$, $d = -100$

🖎*Given a term in an arithmetic sequence and the common difference find the first five terms and the explicit formula.*

5) $a_{36} = -276$, $d = -7$

6) $a_{37} = 249$, $d = 8$

7) $a_{38} = -53.2$, $d = -1.1$

8) $a_{40} = -1191$, $d = -30$

🖎*Given a term in an arithmetic sequence and the common difference find the recursive formula and the three terms in the sequence after the last one given.*

9) $a_{22} = -44$, $d = -2$

10) $a_{12} = 28.6$, $d = 1.8$

11) $a_{18} = 27.4$, $d = 1.1$

12) $a_{21} = -1.4$, $d = 0.6$

Geometric Sequences

🖎 *Determine if the sequence is geometric. If it is, find the common ratio.*

1) $1, -5, 25, -125, \ldots$

3) $4, 16, 36, 64, \ldots$

2) $-2, -4, -8, -16, \ldots$

4) $-3, -15, -75, -375, \ldots$

🖎 *Given the first term and the common ratio of a geometric sequence find the first five terms and the explicit formula.*

5) $a_1 = 0.8, r = -5$

6) $a_1 = 1, r = 2$

🖎 *Given the recursive formula for a geometric sequence find the common ratio, the first five terms, and the explicit formula.*

7) $a_n = a_{n-1} \cdot 2, a_1 = 2$

9) $a_n = a_{n-1} \cdot 5, a_1 = 2$

8) $a_n = a_{n-1} \cdot -3, a_1 = -3$

10) $\quad a_n = a_{n-1} \cdot 3, a_1 = -3$

🖎 *Given two terms in a geometric sequence find the 8th term and the recursive formula.*

11) $\quad a_4 = 12$ and $a_5 = -6$

12) $\quad a_5 = 768$ and $a_2 = 12$

Comparing Arithmetic and Geometric Sequences

✎ *For each sequence, state if it arithmetic, geometric, or neither.*

1) 1, 4, 9, 16, 25, …

2) 1, 5, 25, 125, 625, …

3) 4, 36, 64, 100, …

4) −29, −34, −39, −44, −49, …

5) −4, 12, −36, 108, −324, …

6) 40, 43, 46, 49, 52, …

7) 1, 3, 6, 10, 15, …

8) −34, −26, −18, −10, −2, …

9) $a_n = -163 + 200_n$

10) $a_n = 16 + 3_n$

11) $a_n = -4 \cdot (-3)^{n-1}$

12) $a_n = -43 + 4_n$

13) $a_n = (2n)^2$

14) $a_n = -43 + 7_n$

15) $a_n = -(-3)^{n-1}$

16) $a_n = 2 \cdot (-3)^{n-1}$

Finite Geometric Series

✏️ *Evaluate the related series of each sequence.*

1) $-1, 5, -25, 125$

3) $-1, 4, -16, 64$

2) $-2, 6, -18, 54, -162$

4) $2, 12, 72, 432$

✏️ *Evaluate each geometric series described.*

5) $1 + 2 + 4 + 8 \dots, n = 6$

10) $-3 -6 -12 - 24 \dots, n = 9$

6) $1 - 4 + 16 - 64 \dots, n = 9$

11) $\sum_{n=1}^{8} 2 \cdot (-2)^{n-1}$

7) $-2 - 6 - 18 -54 \dots, n = 9$

12) $\sum_{n=1}^{9} 4 \cdot 3^{n-1}$

8) $2 - 10 + 50 - 250 \dots, n = 8$

13) $\sum_{n=1}^{10} 4 \cdot (-3)^{n-1}$

9) $1 - 5 + 25 - 125 \dots, n = 7$

14) $\sum_{m=1}^{9} -2^{m-1}$

Infinite Geometric Series

✍ *Determine if each geometric series converges or diverges.*

1) $a_1 = -3$, $r = 4$

2) $a_1 = 5.5$, $r = 0.5$

3) $a_1 = -1$, $r = 3$

4) $81 + 27 + 9 + 3 \ldots,$

5) $-3 + \frac{12}{5} - \frac{48}{25} + \frac{192}{125} \ldots,$

6) $\frac{128}{3125} - \frac{64}{625} + \frac{32}{125} - \frac{16}{25} \ldots,$

✍ *Evaluate each infinite geometric series described.*

7) $a_1 = 3$, $r = -\frac{1}{5}$

8) $a_1 = 1$, $r = -3$

9) $a_1 = 1$, $r = -4$

10) $a_1 = 3$, $r = \frac{1}{2}$

11) $1 + 0.5 + 0.25 + 0.125 + \ldots$

12) $81 - 27 + 9 - 3 \ldots,$

13) $1 - 0.6 + 0.36 - 0.216 \ldots,$

14) $3 + \frac{9}{4} + \frac{27}{16} + \frac{81}{64} \ldots,$

15) $\sum_{k=1}^{\infty} 4^{k-1}$

16) $\sum_{i=1}^{\infty} (\frac{1}{3})^{i-1}$

Answers of Worksheets – Chapter 23

Arithmetic Sequences

1) First Five Terms: 24, 26, 28, 30, 32, Explicit: $a_n = 22 + 2n$
2) First Five Terms: −15, −20, −25, −30, −35, Explicit: $a_n = −10 − 5n$
3) First Five Terms: 18, 28, 38, 48, 58, Explicit: $a_n = 8 + 10n$
4) First Five Terms: −38, −138, −238, −338, −438, Explicit: $a_n = 62 − 100n$
5) First Five Terms: −31, −38, −45, −52, −59, Explicit: $a_n = −24 − 7n$
6) First Five Terms: −39, −31, −23, −15, −7, Explicit: $a_n = −47 + 8n$
7) First Five Terms: −12.5, −13.6, −14.7, −15.8, −16.9,   Explicit: $a_n = −11.4 − 1.1n$
8) First Five Terms: −21, −51, −81, −111, −141,   Explicit: $a_n = 9 − 30n$
9) Next 3 terms: −46, −48, −50, Recursive: $a_n = a_{n−1} − 2$, $a_1 = −2$
10) Next 3 terms: 30.4, 32.2, 34, Recursive: $a_n = a_{n−1} + 1.8$, $a_1 = 8.8$
11) Next 3 terms: 28.5, 29.6, 30.7, Recursive: $a_n = a_{n−1} + 1.1$, $a_1 = 8.7$
12) Next 3 terms: −0.8, −0.2, 0.4, Recursive: $a_n = a_{n−1} + 0.6$, $a_1 = −13.4$

Geometric Sequences

1) $r = −5$
2) $r = 2$
3) not geometric
4) $r = 5$
5) First Five Terms: 0.8, −4, 20, −100, 500

   Explicit: $a_n = 0.8 . (−5)^{n−1}$

6) First Five Terms: 1, 2, 4, 8, 16

   Explicit: $a_n = 2^{n−1}$

7) Common Ratio: $r = 2$

   First Five Terms: 2, 4, 8, 16. 32

   Explicit: $a_n = 2 . 2^{n−1}$

8) Common Ratio: $r = −3$

   First Five Terms: −3, 9, −27, 81, −243

   Explicit: $a_n = −3 . (−3)^{n−1}$

9) Common Ratio: r = 5

First Five Terms: 2, 10, 50, 250, 1250

Explicit: $a_n = 2 . 5^{n-1}$

10) Common Ratio: r = 3

First Five Terms: $-3, -9, -27, -81, -243$

Explicit: $a_n = -3 . 3^{n-1}$

11) $a_8 = \frac{3}{4}$, Recursive: $a_n = a_{n-1} . \frac{-1}{2}$, $a_1 = -96$

12) $a_8 = 49152$, Recursive: $a_n = a_{n-1} . 4$, $a_1 = 3$

Comparing Arithmetic and Geometric Sequences

1) Neither
2) Geometric
3) Neither
4) Arithmetic
5) Geometric
6) Arithmetic

7) Neither
8) Arithmetic
9) Arithmetic
10) Arithmetic
11) Geometric
12) Arithmetic

13) Neither
14) Arithmetic
15) Geometric
16) Geometric

Finite Geometric

1) 104
2) −122
3) 51
4) 518
5) 63

6) 52429
7) −19682
8) −130208
9) 13021
10) −1533

11) −170
12) 39364
13) −59048
14) −511

Infinite Geometric

1) Diverges
2) Converges
3) Diverges
4) Converges
5) Converges
6) Diverges

7) $\frac{5}{2}$
8) No sum
9) No sum
10) 6
11) 2

12) $\frac{243}{4}$
13) 0.625
14) 12
15) No sum
16) $\frac{3}{2}$

TSI Mathematics Practice Tests

TSI tests use a multiple-choice format and there's no time limit on the tests, so you can focus on doing your best to demonstrate your skills.

TSI uses the computer-adaptive technology and the questions you see are based on your skill level. Your response to each question drives the difficulty level of the next question.

There are 20 Mathematics questions on TSI test covering three Math sections:

Arithmetic

The Arithmetic test measures your ability to perform basic arithmetic operations and to solve problems that involve fundamental arithmetic concepts.

Elementary Algebra

The Elementary Algebra test measures your ability to perform basic algebraic operations and to solve problems involving elementary algebraic concepts.

College-Level Math

The College-Level Math test measures your ability to solve problems that involve college-level mathematics concepts.

TSI does NOT permit the use of personal calculators on the Math portion of placement test. TSI expects students to be able to answer certain questions without the assistance of a calculator. Therefore, they provide an onscreen calculator for students to use on some questions.

In this section, there are two complete TSI Mathematics Tests. Take these tests to see what score you'll be able to receive on a real TSI test.

Good luck!

Time to Test

Time to refine your skill with a practice examination

Take practice TSI Math Tests to simulate the test day experience. After you've finished, score your tests using the answer keys.

Before You Start

- You'll need a pencil and scratch papers to take the tests.

- For these practice tests, don't time yourself. Spend time as much as you need.

- It's okay to guess. You won't lose any points if you're wrong.

- After you've finished the test, review the answer key to see where you went wrong.

Mathematics is like love; a simple idea, but it can get complicated.

TSI Mathematics Practice

Test 1

(Non–Calculator)

2 Sections – 20 questions

Total time for two sections: No Time Limit

You may not use a calculator on this section.

Arithmetic and Elementary Algebra

1) $(x + 7)(x + 5) =$

$$x^2 + 5x + 7x + 12$$

$$x^2 + 12x + 12$$

A. $x^2 + 12x + 12$

B. $2x + 12x + 12$

C. $x^2 + 35x + 12$

D. $x^2 + 12x + 35$

2) If x is a positive integer divisible by 6, and $x < 60$, what is the greatest possible value of x?

$$x < 60$$

A. 54

B. 48

C. 36

D. 59

3) $x^2 - 81 = 0$, x could equal to:

A. 6

B. 9

C. 12

D. 15

4) If a = 8, what is the value of b in this equation?

$$b = \frac{a^2}{4} + 4$$

A. 24

B. 22

C. 20

D. 28

5) If $6.5 < x \le 9.0$, then x cannot be equal to:

A. 6.5

B. 9

C. 7.2

D. 7.5

6) What is the area of an isosceles right triangle that has one leg that measures 6 cm?

A. 18 cm

B. 36 cm

C. $6\sqrt{2}$ cm

D. 72 cm

7) Which of the following expressions is equivalent to $10 - \frac{2}{3}x \geq 12$

A. $x \geq -3$

B. $x \leq -3$

C. $x \geq 24\frac{1}{3}$

D. $x \leq 24\frac{1}{3}$

8) Which of the following is a factor of both $x^2 - 2x - 8$ and $x^2 - 6x + 8$?

A. $(x - 4)$

B. $(x + 4)$

C. $(x - 2)$

D. $(x + 2)$

9) $\frac{1}{6b^2} + \frac{1}{6b} = \frac{1}{b^2}$, then $b = ?$

A. $-\frac{16}{15}$

B. 5

C. $-\frac{15}{16}$

D. 8

10) If two angles in a triangle measure 53 degrees and 45 degrees, what is the value of the third angle?

A. 8 degrees

B. 42 degrees

C. 82 degrees

D. 98 degrees

College–Level Mathematics

1) If $f(x) = 5 + x$ and $g(x) = -x^2 - 1 - 2x$, then find $(g - f)(x)$?

    A. $x^2 - 3x - 6$

    B. $x^2 - 3x + 6$

    C. $-x^2 - 3x + 6$

    D. $-x^2 - 3x - 6$

2) $\frac{|3+x|}{7} \le 5$, then $x = ?$

    A. $-38 \le x \le 35$

    B. $-38 \le x \le 32$

    C. $-32 \le x \le 38$

    D. $-32 \le x \le 32$

3) $\tan\left(-\frac{\pi}{6}\right) = ?$

    A. $\frac{\sqrt{3}}{3}$

    B. $-\frac{\sqrt{2}}{2}$

    C. $\frac{\sqrt{2}}{2}$

    D. $-\frac{\sqrt{3}}{3}$

4) $\dfrac{\sqrt{32\ ^5b^3}}{\sqrt{2ab^2}}$ = ?

A. $4a^2\sqrt{b}$

B. $2b^2\sqrt{a}$

C. $4b^2\sqrt{a}$

D. $-4a^2\sqrt{b}$

5) The cost, in thousands of dollars, of producing x thousands of textbooks is C $(x) = x^2 +$ 10x + 30. The revenue, also in thousands of dollars, is R(x) = 4x. Find the profit or loss if 3,000 textbooks are produced. (profit = revenue − cost)

A. $21,000 loss

B. $57,000 profit

C. $3,000 profit

D. $57,000 loss

6) Suppose a triangle has the dimensions indicated below:

Then Sin B = ?

A. $\dfrac{3}{5}$

B. $\dfrac{4}{5}$

C. $\dfrac{4}{3}$

D. $\dfrac{3}{4}$

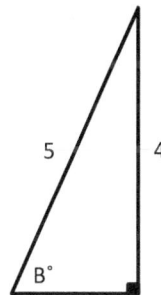

7) Find the slope–intercept form of the graph $6x - 7y = -12$

A. $y = -\dfrac{7}{6}x - \dfrac{12}{7}$

B. $y = -\dfrac{6}{7}x + 12$

C. $y = \dfrac{6}{7}x + \dfrac{12}{7}$

D. $y = \dfrac{7}{6}x - 12$

8) Ella (E) is 4 years older than her friend Ava (A) who is 3 years younger than her sister Sofia (S). If E, A and S denote their ages, which one of the following represents the given information?

A. $\begin{cases} E = A + 4 \\ S = A - 3 \end{cases}$

B. $\begin{cases} E = A + 4 \\ A = S + 3 \end{cases}$

C. $\begin{cases} A = E + 4 \\ S = A - 3 \end{cases}$

D. $\begin{cases} E = A + 4 \\ A = S - 3 \end{cases}$

9) Which of the following point is the solution of the system of equations?

$$\begin{cases} 5x + y = 9 \\ 10x - 7y = -18 \end{cases}$$

A. (2, 4)

B. (2, 2)

C. (1, 4)

D. (0, 4)

10) Find the Center and Radius of the graph $(x - 3)^2 + (y + 6)^2 = 12$

A. (3, 6), $\sqrt{3}$

B. (3, −6), $2\sqrt{3}$

C. (−3, 6), $2\sqrt{3}$

D. (3, −6), $\sqrt{3}$

TSI Mathematics Practice

Test 2

(Non–Calculator)

2 Sections – 20 questions

Total time for two sections: No Time Limit

You may not use a calculator on this section.

Arithmetic and Elementary Algebra

1) $(x - 4)(x^2 + 5x + 4) = ?$

    A. $x^3 + x^2 - 16x + 16$

    B. $x^3 + 2x^2 - 16x - 16$

    C. $x^3 + x^2 - 16x - 16$

    D. $x^3 + x^2 + 16x - 15$

2) How many 3 × 3 squares can fit inside a rectangle with a height of 54 and width of 12?

    A. 72

    B. 52

    C. 62

    D. 42

3) If $7 + 2x \leq 15$, what is the value of $x \leq$?

    A. $14x$

    B. 4

    C. -4

    D. $15x$

4) Liam's average (arithmetic mean) on two mathematics tests is 8. What should Liam's score be on the next test to have an overall of 9 for all the tests?

    A. 8

    B. 9

    C. 10

    D. 11

5) $7^7 \times 7^8 = ?$

    A. 7^{56}

    B. $7^{0.89}$

    C. 7^{15}

    D. 1^7

6) What is 5231.48245 rounded to the nearest tenth?

    A. 5231.482

    B. 5231.5

    C. 5231

    D. 5231.48

7) 15 is what percent of 75?

    A. 10%

    B. 20%

    C. 30%

    D. 40%

8) Last Friday Jacob had $32.52. Over the weekend he received some money for cleaning the attic. He now has $44. How much money did he receive?

    A. $76.52

    B. $11.48

    C. $32.08

    D. $12.58

9) Simplify $\dfrac{\frac{1}{2} - \frac{x+5}{4}}{\frac{x^2}{2} - \frac{5}{2}}$

    A. $\dfrac{3-x}{x^2-10}$

    B. $\dfrac{3-x}{2x^2-10}$

    C. $\dfrac{3+x}{x^2-10}$

    D. $\dfrac{-3-x}{2x^2-10}$

10) $\sqrt{47}$ is between which two whole numbers?

    A. 3 and 4

    B. 4 and 5

    C. 5 and 6

    D. 6 and 7

College–Level Mathematics

1) Solve the equation: $\log_4(x+2) - \log_4(x-2) = 1$

   A. 10

   B. $\dfrac{3}{10}$

   C. $\dfrac{10}{3}$

   D. 3

2) Solve $e^{5x+1} = 10$

   A. $\dfrac{\ln(10) + 1}{5}$

   B. $\dfrac{\ln(10) - 1}{5}$

   C. $5\ln(10) + 2$

   D. $5\ln(10) - 2$

3) If $f(x) = x - \dfrac{5}{3}$ and f^{-1} is the inverse of $f(x)$, what is the value of $f^{-1}(5)$?

   A. $\dfrac{10}{3}$

   B. $\dfrac{3}{20}$

   C. $\dfrac{20}{3}$

   D. $\dfrac{3}{10}$

4) What is cos 30∘?

    A. $\frac{1}{2}$

    B. $\frac{\sqrt{2}}{2}$

    C. $\frac{\sqrt{3}}{2}$

    D. $\sqrt{3}$

5) If θ is an acute angle and sin θ = $\frac{3}{5}$, then cos θ = ?

    A. −1

    B. 0

    C. $\frac{4}{5}$

    D. $\frac{5}{4}$

6) What is the solution of the following system of equations?

$$\begin{cases} -2x - y = -9 \\ 5x - 2y = 18 \end{cases}$$

    A. (−1, 2)

    B. (4, 1)

    C. (1, 4)

    D. (4, −2)

7) Solve.

$|9 - (12 \div | \, 2 - 5 \, |)| = ?$

   A. 9

   B. −6

   C. 5

   D. −5

8) If $\log_2 x = 5$, then $x = ?$

   A. 2^{10}

   B. $\dfrac{5}{2}$

   C. 2^6

   D. 32

9) What's the reciprocal of $\dfrac{x^3}{16}$?

   A. $\dfrac{16}{x^3} - 1$

   B. $\dfrac{48}{x^3}$

   C. $\dfrac{16}{x^3} + 1$

   D. $\dfrac{16}{x^3}$

10) Find the inverse function for ln $(2x + 1)$

A. $\frac{1}{2}(e^x - 1)$

B. $(e^x + 1)$

C. $\frac{1}{2}(e^x + 1)$

D. $(e^x - 1)$

www.EffortlessMath.com

TSI Math Practice Test 1 Answer Key

✳Now, it's time to review your results to see where you went wrong and what areas you need to improve!

Arithmetic and Elementary Algebra

1.	D		2.	A
3.	B		4.	C
5.	A		6.	A
7.	B		8.	A
9.	B		10.	C

College–Level Mathematics Test

1.	D		2.	B
3.	D		4.	A
5.	D		6.	B
7.	C		8.	D
9.	C		10.	B

TSI Math Practice Test 2 Answer Key

Arithmetic and Elementary Algebra

1.	C		2.	A
3.	B		4.	D
5.	C		6.	B
7.	B		8.	B
9.	D		10.	D

College–Level Mathematics Test

1.	C		2.	B
3.	C		4.	C
5.	C		6.	B
7.	C		8.	D
9.	D		10.	A

TSI Mathematics Practice Test 1 Answers and Explanations

Arithmetic and Elementary Algebra

1) Choice D is correct

Use FOIL (First, Out, In, Last)

$(x + 7)(x + 5) = x^2 + 5x + 7x + 35 = x^2 + 12x + 35$

2) Choice A is correct

$\frac{54}{6} = \frac{27}{3} = 9$, $\frac{48}{6} = \frac{24}{3} = 8$, $\frac{36}{6} = \frac{18}{3} = 6$, $\frac{59}{6} = \frac{59}{6}$ 59 is prime number

3) Choice B is correct

$x^2 - 81 = 0 \Rightarrow x^2 = 81 \Rightarrow x = 9$

4) Choice C is correct

If $a = 8$ then $b = \frac{8^2}{4} + 4 \Rightarrow b = \frac{64}{4} + 4 \Rightarrow b = 16 + 4 = 20$

5) Choice A is correct

If $6.5 < x \leq 9.0$, then x cannot be equal to 6.5

6) Choice A is correct

$a = 6 \Rightarrow$ area of triangle is $= \frac{1}{2}(6 \times 6) = \frac{36}{2} = 18$ cm

Isosceles right triangle

7) Choice B is correct

Simplify:

$10 - \frac{2}{3}x \geq 12 \Rightarrow -\frac{2}{3}x \geq 2 \Rightarrow -x \geq 3 \Rightarrow x \leq -3$

8) Choice A is correct

Factor each trinomial $x^2 - 2x - 8$ and $x^2 - 6x + 8$

$x^2 - 2x - 8 \Rightarrow (x - 4)(x + 2)$

$x^2 - 6x + 8 \Rightarrow (x - 2)(x - 4)$

9) Choice B is correct

$\frac{1 + b}{6b^2} = \frac{1}{b^2} \Rightarrow (b \neq 0)\ b^2 + b^3 = 6b^2 \Rightarrow b^3 - 5b^2 = 0 \Rightarrow b^2(b - 5) = 0 \Rightarrow b - 5 = 0 \Rightarrow b = 5$

10) Choice C is correct

$53° + 45° = 98°$

$180° - 98° = 82°$

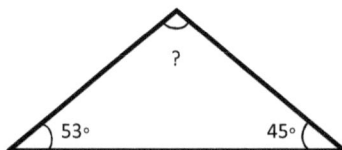

The value of the third angle is $82°$.

College–Level Mathematics

1) **Choice D is correct**

$(g - f)(x) = g(x) - f(x) = (-x^2 - 1 - 2x) - (5 + x)$

$-x^2 - 1 - 2x - 5 - x = -x^2 - 3x - 6$

2) **Choice B is correct**

$\dfrac{|3 + x|}{7} \leq 5 \Rightarrow |3 + x| \leq 35 \Rightarrow -35 \leq 3 + x \leq 35 \Rightarrow -35 - 3 \leq x \leq 35 - 3 \Rightarrow$

$-38 \leq x \leq 32$

3) **Choice D is correct**

$\tan\left(-\dfrac{\pi}{6}\right) = -\dfrac{\sqrt{3}}{3}$

4) **Choice A is correct**

$\dfrac{\sqrt{32a^5b^3}}{\sqrt{2ab^2}} = \dfrac{4a^2b\sqrt{2ab}}{b\sqrt{2a}} = 4a^2\sqrt{b}$

5) **Choice D is correct**

$c(3) = (3)^2 + 10(3) + 30 = 9 + 30 + 30 = 69$

$4 \times 3 = 12 \Rightarrow 12 - 69 = -57 \Rightarrow 57{,}000 \text{ loss}$

6) **Choice B is correct**

$\sin B = \dfrac{\text{the length of the side that is opposite that angle}}{\text{the length of the longest side of the triangle}} = \dfrac{4}{5}$

7) **Choice C is correct**

$-7y = -6x - 12 \Rightarrow y = \dfrac{-6}{-7}x - \dfrac{12}{-7} \Rightarrow y = \dfrac{6}{7}x + \dfrac{12}{7}$

8) Choice D is correct

$E = 4 + A$

$A = S - 3$

9) Choice C is correct

$$\begin{cases} 5x + y = 9 \\ 10x - 7y = -18 \end{cases} \Rightarrow \text{Multiplication } (-2) \text{ in first equation} \Rightarrow \begin{cases} -10x - 2y = -18 \\ 10x - 7y = -18 \end{cases}$$

Add two equations together $\Rightarrow -9y = -36 \Rightarrow y = 4$ then: $x = 1$

10) Choice B is correct

$(x - h)^2 + (y - k)^2 = r^2 \quad \Rightarrow$ center: (h, k) and radius: r

$(x - 3)^2 + (y + 6)^2 = 12 \Rightarrow$ center: $(3, -6)$ and radius: $2\sqrt{3}$

TSI Mathematics Practice Test 2 Answers and Explanations

Arithmetic and Elementary Algebra

1) Choice C is correct

Use FOIL (First, Out, In, Last)

$(x - 4)(x^2 + 5x + 4) = x^3 + 5x^2 + 4x - 4x^2 - 20x - 16$

$= x^3 + x^2 - 16x - 16$

2) Choice A is correct

Number of squares equal to: $\frac{54 \times 12}{3 \times 3} = 18 \times 4 = 72$

3) Choice B is correct

Simplify:

$7 + 2x \leq 15 \Rightarrow 2x \leq 15 - 7 \Rightarrow 2x \leq 8 \Rightarrow x \leq 4$

4) Choice D is correct

$\frac{a + b}{2} = 8 \qquad \Rightarrow \qquad a + b = 16$

$\frac{a + b + c}{3} = 9 \qquad \Rightarrow \qquad a + b + c = 27$

$16 + c = 27 \qquad \Rightarrow \qquad c = 27 - 16 = 11$

5) Choice C is correct

$7^7 \times 7^8 = 7^{7+8} = 7^{15}$

6) Choice B is correct

Underline the tenth place:

5231.$\underline{4}$8245

Look to the right if it is 5 or above, give it a shove.

Then, round up to 5231.5

7) Choice B is correct

$$75 \times \frac{x}{100} = 15 \qquad \Rightarrow \qquad 75 \times x = 1500 \qquad \Rightarrow \qquad x = \frac{1500}{75} = 20$$

8) Choice B is correct

$44 - $32.52 = $11.48

9) Choice D is correct

Simplify:

$$\frac{\frac{1}{2} - \frac{x+5}{4}}{\frac{x^2}{2} - \frac{5}{2}} = \frac{\frac{1}{2} - \frac{x+5}{4}}{\frac{x^2 - 5}{2}} = \frac{2(\frac{1}{2} - \frac{x+5}{4})}{x^2 - 5}$$

$\Rightarrow$ Simplify: $\dfrac{1}{2} - \dfrac{x+5}{4} = \dfrac{-x-3}{4}$

then: $\dfrac{2(\frac{-x-3}{4})}{x^2 - 5} = \dfrac{\frac{-x-3}{2}}{x^2 - 5} = \dfrac{-x-3}{2(x^2 - 5)} = \dfrac{-x-3}{2x^2 - 10}$

10) Choice D is correct

$\sqrt{47}$ = 6.85565...

then: $\sqrt{47}$ is between 6 and 7

College–Level Mathematics Test

1) Choice C is correct

<u>METHOD ONE</u>

$\log_4(x + 2) - \log_4(x - 2) = 1$

Add $\log_4(x - 2)$ to both sides

$\log_4(x + 2) - \log_4(x - 2) + \log_4(x - 2) = 1 + \log_4(x - 2)$

$\log_4(x + 2) = 1 + \log_4(x - 2)$

Apply logarithm rule: $a = \log_b(b^a) \Rightarrow 1 = \log_4(4^1) = \log_4(4)$

then: $\log_4(x + 2) = \log_4(4) + \log_4(x - 2)$

Logarithm rule: $\log_c(a) + \log_c(b) = \log_c(ab)$

then: $\log_4(4) + \log_4(x - 2) = \log_4(4(x - 2))$

$\log_4(x + 2) = \log_4(4(x - 2))$

When the logs have the same base: $\log_b(f(x)) = \log_b(g(x)) \Rightarrow f(x) = g(x)$

$(x + 2) = 4(x - 2)$

$x = \dfrac{10}{3}$

<u>METHOD TWO</u>

We know that: $\quad\log_a b - \log_a c = \log_a \dfrac{b}{c} \quad$ and $\quad \log_a b = c \Rightarrow b = a^c$

Then: $\log_4(x + 2) - \log_4(x - 2) = \log_4 \dfrac{x+2}{x-2} = 1 \Rightarrow \dfrac{x+2}{x-2} = 4^1 = 4 \Rightarrow x + 2 = 4(x - 2)$

$\Rightarrow x + 2 = 4x - 8 \Rightarrow 4x - x = 8 + 2 \rightarrow 3x = 10 \Rightarrow x = \dfrac{10}{3}$

2) Choice B is correct

$e^{5x + 1} = 10$

If $f(x) = g(x)$, then $\ln(f(x)) = \ln(g(x))$

$\ln(e^{5x + 1}) = \ln(10)$

Apply logarithm rule: $\log_a(x^b) = b \log_a(x)$

$\ln(e^{5x + 1}) = (5x + 1)\ln(e)$

$(5x + 1)\ln(e) = \ln(10)$

$(5x + 1)\ln(e) = (5x + 1)$

$(5x + 1) = \ln(10) \qquad\qquad \Rightarrow \qquad x = \dfrac{\ln(10) - 1}{5}$

3) Choice C is correct

$f(x) = x - \dfrac{5}{3} \quad \Rightarrow \quad y = x - \dfrac{5}{3} \Rightarrow \quad y + \dfrac{5}{3} = x$

$f^{-1} = x + \dfrac{5}{3}$

$f^{-1}(5) = 5 + \dfrac{5}{3} = \dfrac{20}{3}$

4) Choice C is correct

$\cos 30° = \dfrac{\sqrt{3}}{2}$

5) Choice C is correct

$\sin\theta = \dfrac{3}{5} \Rightarrow$ we have following triangle, then

$c = \sqrt{5^2 - 3^2} = \sqrt{25 - 9} = \sqrt{16} = 4$

$\cos\theta = \dfrac{4}{5}$

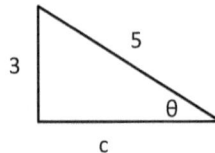

6) Choice B is correct

$\begin{cases} -2x - y = -9 \\ 5x - 2y = 18 \end{cases}$ ⇒ Multiplication (–2) in first equation ⇒ $\begin{cases} 4x + 2y = 18 \\ 5x - 2y = 18 \end{cases}$

Add two equations together ⇒ $9x = 36 \Rightarrow x = 4$ then: $y = 1$

7) Choice C is correct

$|9 - (12 \div | 2 - 5 |)| = |9 - (12 \div |-3|)| = |9 - (12 \div 3)| = |9 - 4| = |5| = 5$

8) Choice D is correct

METHOD ONE

$\log_2 x = 5$

Apply logarithm rule: $a = \log_b(b^a)$

$5 = \log_2(2^5) = \log_2(32)$

$\log_2 x = \log_2(32)$

When the logs have the same base: $\log_b(f(x)) = \log_b(g(x)) \Rightarrow f(x) = g(x)$

then: $x = 32$

METHOD TWO

We know that: $\log_a b = c \Rightarrow b = a^c$ $\log_2 x = 5 \Rightarrow x = 2^5 = 32$

9) Choice D is correct

$\dfrac{x^3}{16}$ ⇒ reciprocal is : $\dfrac{16}{x^3}$

10) Choice A is correct

$f(x) = \ln(2x + 1)$

$y = \ln(2x + 1)$

Change variables x and y: $x = \ln(2y + 1)$

solve: $x = \ln(2y + 1)$

$y = \dfrac{e^x - 1}{2} = \dfrac{1}{2}(e^x - 1)$

"Effortless Math Education" Publications

Effortless Math authors' team strives to prepare and publish the best quality TSI Mathematics learning resources to make learning Math easier for all. We hope that our publications help you learn Math in an effective way and prepare for the TSI test.

We all in Effortless Math wish you good luck and successful studies!

Effortless Math Authors

Online Math Lessons

Enjoy interactive Math lessons online

with the best Math teachers

Online Math learning that's effective, affordable, flexible, and fun

Learn Math wherever you want; when you want

Ultimate flexibility. You can now learn Math online, enjoy high quality engaging lessons no matter where in the world you are. It's affordable too.

Learn Math with one-on-one classes

We provide one-on-one Math tutoring online. We believe that one-to-one tutoring is the most effective way to learn Math.

Qualified Math tutors

Working with the best Math tutors in the world is the key to success! Our tutors give you the support and motivation you need to succeed with a personal touch.

Online Math Lessons

It's easy! Here's how it works.

1- Request a FREE introductory session.

2- Meet a Math tutor online.

3- Start Learning Math in Minutes.

Send Email to: **info@EffortlessMath.com**

CPSIA information can be obtained
at www.ICGtesting.com
Printed in the USA
LVHW061708241218
601588LV00024B/311/P

9 781984 924544